HOW TO THRIVE IN THE NEXT ECONOMY

John Thackara

HOW TO THRIVE IN
THE NEXT ECONOMY
DESIGNING TOMORROW'S WORLD TODAY

Thames & Hudson

FOR LEX AND ELEANOR THACKARA

First published in 2015 in hardcover
in the United States of America by
Thames & Hudson Inc., 500 Fifth Avenue,
New York, New York 10110

thamesandhudsonusa.com

Library of Congress Catalog Card Number
2015938380

ISBN 978-0-500-51808-3

Printed and bound in India by Replika Press Pvt. Ltd.

CONTENTS

CHANGING:
FROM DO LESS HARM,
TO LEAVE THINGS BETTER

At a dusty crossing on the long cross-country road from Kanpur to Lucknow, in Uttar Pradesh, India, we come across a huge video screen on the back of a flat-bed truck. Together with a dozen villagers, four people on bicycles, and a cow, we stare in a daze at the screen. On the left side of the screen the landscape on each side of the River Ganges, in whose vast fertile plain we are standing, is made to look hot, dusty, and wretched. On the right of the screen, a better future is portrayed: busy cities, robot assembly lines, and high-speed trains. This before-and-after sequence is followed by a full-screen video in which computer-generated apartment blocks sprout like so many mushrooms from bright green grass along the banks of the River Ganges. 'Welcome to Trans-Ganga HighTech City,' explains the voiceover.

'May the odds be ever in your favour!' mutters my young companion. 'This is pure *Hunger Games*,' she explains, and goes on to describe how, in a film that everyone in the world has seen except me, a young woman called Katniss lives in a dystopian, post-apocalyptic nation. Every year The Capitol, where the rich people live, asserts its power over the poor regions that surround it by staging the Hunger Games in which boys and girls, selected by lottery from the poor areas, compete in a televised battle to the death. 'May the odds be ever in your favour!', I learn, is what the creepy ruler guy says when opening the Games – in which all but one competitor will die.

Trans-Ganga HighTech City resembles *The Hunger Games* all too well – a glossy, gated city surrounded by social hardship and degraded landscapes. Trans-Ganga is one of 100 Indian turn-key cities that developers want to build on green land swept clean of its small farmers and biodiversity. Investors are promised that special laws will be passed to ensure that millions of poor Indians are 'excluded from the privileges of such great infrastructure'.[1] These physical and social impacts are disturbing enough – but what really cranks up the anxiety level are the bright and perky voices, on screens everywhere, proclaiming these developments to be for the good of all. Whenever a voice is raised in protest at the negative impacts of these plans, the perky heads blame the losers for their own misfortune: Get a job! Try harder! May the odds be ever in your favour!

The words we choose are important as we try to make sense of these new times. One man's *energy descent* is another woman's *energy transition*. Talk of an *impending crisis* is scary; realizing that the crisis is already underway, less so. The *end of growth* sounds grim – but it is not the *end of life*. The *collapse of civilization* is a terrifying prospect; *the birth of a new one* puts things in a different light. 'What is civilizational collapse, after all,' quips the Italian physicist Ugo Bardi, a self-styled 'stoic scientist', 'other than a period in which things are changing faster than usual?'[2]

The apocalyptic view is couched in the language of danger and collapse. Industrial civilization has started to crash, say the 'doomers'. For them, our best course of action is to head for the hills with a truckload of guns and peanut butter. At the other extreme, optimistic technology buffs are confident that man-made solutions will soon allow us to carry on as usual. And what about the rest of us? Most people I know are anxious about what's happening around them, but silently so; they think less about the collapse of civilizations than with finding work, or feeding their kids. But they – we – feel less and less secure. It doesn't help that the media are filled with fatuous advice about what we should do: drive a Tesla? Change a light bulb? Give us a break.

This book is that break. It tells of a third social movement – much bigger than the rifle-packing doomers and the green-tech dreamers – that's emerging as the global crisis unfolds. This movement is below the radar of mainstream media, but it contains a million active groups – and rising. Quietly, for the most part, communities the world over are growing a replacement economy from the ground up. As you will read in the pages that follow, their number includes energy angels, wind wizards, and watershed managers. There are bioregional planners, ecological historians, and citizen foresters. Alongside dam removers, river restorers, and rain harvesters, there are urban farmers, seed bankers, and master conservers. You'll meet building dismantlers, office-block refurbishers, and barn raisers. There are natural painters, and green plumbers. There are trailer-park renewers, and land-share brokers. The movement involves computer recyclers, hardware re-mixers, and textile upcyclers. It extends to local currency designers. There are community doctors. And elder carers. And ecological teachers.

For most of the people I write about in this book, the changes they are making are driven by necessity; they are not a lifestyle choice. Few of them are fighting directly for political power, or standing for election. They cluster, instead, under the umbrella of a social and solidarity economy. Different groups and movements have names like Transition Towns, Shareable, Peer to Peer, Degrowth, or *Buen Vivir*. Their number includes FabLabs, hacker spaces, and the maker movement. Some have taken over neglected buildings – from castles and car parks, to ports, piers, hospitals, and former military sites. There are campaigning organizations, too – for slow food, the rights of nature, and seed saving – not to mention bioregionalism, and commoning. And our number is growing. Up to 12 per cent of economically active citizens in Sweden, Belgium, France, Holland, and Italy work in some kind of social enterprise – and that's in addition to the vast amounts of unpaid work already being done in the household and caring economy.

Although these projects are wondrously diverse they are all, for the Spanish writer Amador Fernández-Savater, 'message-bearers of a new story of the world'.[3] A green thread runs through this story: a growing recognition that our lives are codependent with the plants, animals, air, water, and soils that surround us. The philosopher Joanna Macy describes the appearance of this new story as the 'Great Turning' – a profound shift in our perception of who we are, and a reawakening to the fact that we are not separate from the Earth as a complex of living systems.[4] From sub-microscopic viruses, to the vast subsoil networks that support trees, this new story goes, the entire Earth is animated by complex interactions between its life forms, rocks, atmosphere, and water. Explained in this way – by science, as much as by philosophy – the Earth no longer looks like a repository of inert resources. On the contrary: healthy soils, living systems, and the ways we can help them regenerate supply the 'why' of economic activity that's missing from the mainstream story. The one kind of growth that makes sense, in this new story, is the regeneration of life on Earth.

The notion of a living economy can sound poetic, but vague. Where, you may ask, is its manifesto? Who is in charge? These are old-fashioned questions. The account given by Macy – of a quietly unfolding transformation – is consistent with the way scientists, too, explain how complex systems change. By their account, a variety of changes, interventions, and disruptions accumulate across time until the system reaches a tipping point: then, at a moment that cannot be predicted, a small release of energy triggers a much larger release, or phase shift, and the system as a whole transforms. Sustainability, in other words, is not something to be engineered, or demanded from politicians; it's a condition that emerges through incremental as well as abrupt change at many different scales. 'All the great transformations have been unthinkable until they actually came to pass,' confirms the French philosopher Edgar Morin. 'The fact that a belief system is deeply rooted does not mean it cannot change.'[5]

So this is an optimistic book – but not dreamily so. If I'm to convince you that the stories to come are the harbinger of the new economy we so desperately need, I first need to explore the powerful but hidden reasons that a return to normal is just not going to happen.

ENERGY

In 1971 a geologist called Earl Cook evaluated the amount of energy 'captured from the environment' in different economic systems.[6] Cook discovered that a modern city dweller needed about 230,000 kilocalories per day to keep body and soul together. This compared starkly to a hunter-gatherer, ten thousand years earlier, who needed about 5,000 kcal per day to get by. That gap, between simple and complex lives, has widened at an accelerating rate since 1971. Once all the systems, networks, and gadgets of modern life are factored in – the cars, planes, factories, buildings, infrastructure, heating, cooling, lighting, food, water, hospitals, information systems, and their attendant gadgets – well, a New Yorker or Londoner today 'needs' about *sixty times* more energy and resources per person than a hunter-gatherer. To put it another way: American citizens today use more energy and physical resources in a month than our great-grandparents used during their whole lifetime.

This upwards trajectory would be alarming if we thought about it clearly – but we don't. We simply ignore the fact that all these 'needs' depend on growing flows of cheap and intense energy. Belief is one thing; basic mathematics, and the laws of physics, suggest otherwise. The exponential growth of anything tangible, or energy consuming, cannot continue indefinitely in a finite universe. As Tom Murphy, an American physics professor, patiently explains, even if the future rate of compound energy growth in our economy declined to a lower level than today, we'd still see an increase by a factor of 10 every 100 years; in 275 years, we'd reach 600 times our current rates of use. Surely, you may counter, economic growth could be decoupled from energy growth and be freed to expand to infinity that way? Well, no. Multiplying money *always* expands an economy's physical impacts on the Earth. 'Energy is the capacity to do work; it's the lifeblood of activity,' explains Professor Murphy. 'Think it through: to keep GDP growing indefinitely on a fixed energy diet would mean that anything requiring energy becomes an ever-smaller part of GDP, until it carries negligible value. But food, heat, and clothing will never be negligible needs. There is plenty of scope for economic activities that use *less* energy – but that is not the same as reducing energy intensity to zero.'[7] Indefinite GDP growth is Not Going to Happen.

The world is not in danger of running completely out of energy – in the short or even medium term. Strictly speaking, we don't face an energy crisis so much as an *exergy* crisis – that is, a shortage of energy that is so highly concentrated, and easy to obtain, that it can easily be used to drive the economy. At its most dynamic, the thermo-industrial economy grew using oil that, if it did not literally gush out of the ground, was easily extracted using oil-powered machines. Since then, we've burned our way through the easy-to-access fuels and extracting energy gets harder and more expensive every year. To make matters worse, the man-made world has become so much more complicated – think of all those computer networks, aviation systems, and fancy hospitals – that it now takes far more energy just to keep 'the system' going than would have been needed, just a generation ago, to deliver a simple but effective product or service to you and me.

For an explanation of where these trends are taking us I went to the Houses of Parliament in London. An American ecology professor, Charles Hall, had been invited to give a lecture on Energy Return on Energy Invested (EROEI). The central principle of EROEI, he explained, is that it takes energy to obtain energy – and if that process takes too much effort, and therefore cost, then the needed investment probably won't be made – and the energy needed to run the system won't be available. Prof Hall showed us the change, through time, of the number of barrels of oil obtained for use in the economy for every barrel invested in extraction:

1930s 100 barrels for economic use
1970s 25 barrels
1990s 15 barrels

Most of the energy solutions being trumpeted today, Hall continued – from tar sands in Alberta to solar arrays in Spain – fall well below the 15:1 threshold below which the investment never pays for itself. 'You can't have an economy without energy. Energy does the work!' Professor Hall concluded, echoing Tom Murphy's 'poor-quality fuels mean poor-quality growth'. I'll never forget the silence that followed his presentation. Eventually, a senior Member of Parliament stood up, thanked Professor Hall for his 'most interesting presentation', and added, 'but of course, for an elected politician, reduced affluence is an impossible sell'. He then sat down. Professor Hall, the scientist, said he was a numbers guy, not a policy guy – and he sat down, too. Then everybody went home.

Technology optimists believe that renewable energy, conjured into existence by innovation, will allow us to carry on as usual – but they are in for a disappointment. Nearly all plans for a transition to renewable energy suffer from an existential flaw: they take global energy 'needs' as a given, calculate the quantity of renewable energy sources needed to meet them, and then – well, things get vague. Green energy optimists have no answer for a logical inconvenience: it takes astronomical amounts of fossil-fuel energy, and money, to deploy 'green' energy systems – 200 km (125 miles) of copper in one wind turbine, to give just one example. There would be far fewer wind turbines, for example, if they had to be manufactured, installed, and maintained using wind energy. Retrofitting energy systems on a large enough scale to run today's industrial society would require vast investment of materials, money, and organizational effort that, in today's deflationary global crisis, will not be available. Gail Tverberg, an actuary and blogger, puts it bluntly: 'Quite apart from the math, or the thermodynamics, or the simple logic, a lack of cash flow for investment in infrastructure will eventually bring the system down.'[8]

Measured against the laws of mathematics, physics, and common sense, our belief in an energy-intense economy – one that expands to infinity in a finite world – seems irrational. A better word would be *habitual*. Many smart people believe that growth will go on forever because that is all they have known in their lives. They believe in the inevitability of progress because, in their lives at least, things have always progressed. They believe that bold actions should be taken without regard for consequences because there haven't been any negative consequences – or rather, none that they have experienced personally. They believe that man is special, and that progress is unstoppable, because no experience has given them reason to think otherwise. These foundation myths of the modern age – reason, progress, mastery over nature – are oil-powered narratives. In the 1950s, when Milton Friedman expounded the economic thinking that dominates political discourse to this day, you could buy a barrel of oil for US$3.50.

MONEY

The timing and severity of peak energy is a contested topic, but a growing number of people are happy to blame bankers for our economic woes. This blame is misdirected. The men and women in suits can be hard to love, it's true, but they are more the prisoners of a dysfunctional system than its masters – in their case, the money one. And the fate of the money system, it turns out, is tied intimately to the fate of the energy one. Money and energy are better thought of as one story.

Before writing this book, I had vaguely assumed that what banks do is collect deposits and savings from one lot of people, and lend those funds out to different people in the form of loans, mortgages, and credit on plastic cards. This is not the case at all. Although bankers describe their core business as 'lending' money, it should really be described as *creating* money. When you or I borrow money from a bank, and the bank tells you it is 'transferring' funds into your account, that money is not taken out of a vault, nor even sent down a wire from somewhere else. It is newly created, there and then. Only a small fraction of the money they create is backed up by assets – such as the deeds to a house, or a bar of gold – lodged safely in their vaults. For the most part, they just make the loan at will. And it gets curiouser. Even though you and I now have new money to spend, these loans are recorded on the banks' balance sheets as assets. The rationale seems to be that the interest on the loan that you and I will pay to the bank represents a steady flow of profit to them. And because many bankers are paid by commissions on new loans issued, there's a built-in incentive to lend as much as possible.

When an economy is growing, this peculiar dynamic does not much matter: as people buy more goods, often using credit from the bank, and as businesses take out loans to increase their production of goods, then interest on existing loans is repaid. But when economic growth stalls – for example, because there is less cheap energy to power growth – new money stops entering the system and a destructive feedback loop kicks in. Interest on existing loans is not paid; defaults multiply; jobs are lost; people spend less money; businesses take out fewer loans; less new money enters the economy – and the crisis of debt intensifies.

This through-the-looking-glass logic of the money system is made harder to grasp by the ineffable numbers used to describe it. At the time of writing, global debt is estimated to be about US$200 trillion – but what does such a number mean? Well, think of it this way: imagine that a world government, burdened with this debt of $200 trillion, decided to pay it all back at a rate of $1 per second. To pay back $1 million at such a rate would take 11.5 days; paying back $1 billion would take 32 years; but to pay back the full $200 trillion, at a rate of $1 per second, would take 6.5 million years.[9] Taken together with the energy crunch this is why, however much we might wish things to carry on as before, they won't. They can't. As explained by Gail Tverberg, 'An infinite economic growth model has created the need to keep the hamster wheel turning faster and faster until the hamster dies.'[10] Blaming bankers for the hamster's imminent demise is therefore to miss the point; it's the money-energy system itself that is spinning the wheel.

GROWTH

If the manic striving for growth was just about numbers, we could write it off as deluded, but harmless, thinking. But money is not just an abstraction. As professors Murphy and Hall explained above, money gets work done in the real world. When a system must grow in order to survive, but the work it enables is destructive, the consequences are catastrophic.

I experienced the grim consequence of growth for its own sake at a meeting of 200 sustainability managers at a famous home furnishing giant in Sweden. During twenty years of hard work on sustainability, this company has made thousands of rigorously tested improvements; these are recorded on a 'list without end'. The range of improvements is startling – even admirable – except for one fact: the one thing this huge company has *not* done is question whether it should grow. On the contrary: it is committed to double in size by 2020. By that date, the number of customers visiting its giant sheds will increase from 650 million a year at the time of writing to *1.5 billion* a year. And why? The senior manager who briefed our meeting on this plan put this growth into context: 'Growth is needed', she explained, 'to finance the sustainability improvements we all want to make.'

A fatal flaw with this argument is best explained if I talk about wood. The company, as the third largest user of wood in the world, has promised that by 2017 half of all the wood it uses – up from 17 per cent now – will either be recycled or come from forests that are responsibly managed. Now 50 per cent is a vast improvement on 17, but it also begs the question: what about the *other* half of all that wood? As the company doubles in size, that second pile of wood – the *un*-certified half, the unreliably-sourced-at-best half – will soon be twice as big as *all* the wood it uses today. The impact on the world's forests, of this one company's hunger for resources, will be appalling. The committed and gifted people I met in Sweden – along with sustainability teams in hundreds of the world's major companies – are confronted by an awful dilemma: however hard they work, however many leaks they plug in production cycles, the net negative impact of their firm's activities on the world's living systems will be greater in the years ahead than it is today. And all because of compound growth. It doesn't matter how many brands proclaim that their products are verified, accredited, or certified as being sustainable; so long as growth remains a company's prime directive, any promise to leave the world 'as unspoilt as possible' will remain an empty one.

If a lack of data were the main problem, help would be at hand. Following a large international effort, a set of accounting tools known as The Economics of Ecosystems and Biodiversity (TEEB) puts a price on the services provided to industry by nature; and many governments and companies have signed up to its framework.[11] Unfortunately, TEEB has only made things worse. The theory was that knowing the value of ecosystems would cause companies to look after them better – but TEEB's numbers, acting like blood in the water, have also attracted the attention of predatory investors. Living systems – watersheds, minerals, food, and land – are now being converted into 'financialized' assets which, having been rendered abstract, have become new tokens for speculation.[12] By design, these financial products contain powerful incentives for their owners to 'sweat' the underlying assets at an accelerating rate. This commodification of nature has spawned a related but no less baleful phenomenon called 'biodiversity offsetting'. This is the idea that the destruction of an ecosystem by mining, greenfield development, or a large infrastructure project can be 'offset' by the creation of a new patch of nature somewhere else.[13] This scheme is great for the companies digging the mines or pouring the concrete; it also creates new work for an army of intermediaries; but the result on the ground is an acceleration of environmental destruction. Nature is unique and complex. Some ecosystems take hundreds of years to reach their current state. The promise that the habitat can be recreated at will is another false one.[14]

RISK

None of the grim trends I've described above is doomer speculation. Lloyds of London, the epicentre of global risk management, has warned that 'an oil supply crunch is likely in the short-to-medium term'. For another capitalist hotspot, the World Economic Forum (WEF), peak oil is just one item in a guide to possible futures called 'Seeds of Dystopia'. Highlights of this jolly survey include a killer virus pandemic; unmanageable deflation; a geomagnetic storm that wipes out the internet; global food shortages; and 'unprecedented geophysical destruction'.[15] These top-trending risks, says the WEF, 'are a health warning regarding our most critical systems'. The WEF is not alone in its sombre outlook. What its *Global Risks* does for the economy, *Global Trends 2030* does for geopolitics and security.[16] The latter report, published by the US National Intelligence Council, warns that 'we are at a critical juncture in human history…natural disasters might cause governments to collapse'. Climate scientists and ecologists reinforce these warnings. The Stockholm Resilience Centre (SRC), for example, has delineated nine 'planetary boundaries' – the limits, for essential planetary living systems, beyond which we must not go.[17] The SRC's map is alarming enough – we are already beyond the red line on three of its nine systems – but it only plots the *known* risks. Even more alarming is the possibility of a so-called 'ecological surprise'– a transformational change, in one or more natural or man-made systems, that could be sudden, non-linear, and catastrophic. As complex systems researcher Noah Raford explains it, too much interconnectivity makes systems vulnerable to 'phase transition' – a word that sounds more benign than it probably is. When a system reaches a critical state, Raford explains, 'even a tiny change can lead to massive fluctuation and collapse'.[18] We know these events can happen, but we don't know when; they cannot be predicted.

All this is meat and drink to the doomer community – but not everyone agrees that we need to take these risks seriously. On the contrary: trends that signal 'risk' to you or me are embraced by others as opportunities. For some tech boosters, the increased interdependency of systems is good news; it signifies that our economy is in an 'evolutionary uplift' towards a 'post-productive' mode.[19] Boundaries and limits are also anathema to the WEF; risks are described in its doom-filled reports as 'transformational opportunities' that we should grab with relish to 'improve the state of the world' and to pursue the 'critical goal…of future growth'.[20] There is no acknowledgment – not a word – that compound economic growth could possibly be the cause of these biosphere-threatening trends.

As for the fact that exponential economic growth on a physical planet contravenes basic laws of physics and mathematics[21] – that, too, is simply ignored.

This is not to deny that resilience – 'the capacity to bounce back' as one book[22] so well explains it – is a desirable condition. The trouble is that a lot of people perceive resilience – dynamic or otherwise – to be a new variety of risk management that affords them the opportunity to carry on with business as usual. 'We can't avoid shocks in an increasingly complex world,' said one commentator, 'we can only build better shock absorbers.' This metaphor would just about work if the world around us were indeed a tarmac road disfigured by potholes – but it's not. Those 'bumps' we're driving over are better understood as the bodies, metaphorical or otherwise, of living systems.

METABOLIC RIFT

Why would anyone even *consider* driving over them? These powerful individuals are not stupid – so why do they believe so strongly in an ecocidal system? The explanation that works best for me is the existence of a 'metabolic rift' between man and the Earth. This is the idea that a combination of paved surfaces and pervasive media have rendered us cognitively blind to the health of the living systems of which we are a part.[23] As Timothy Morton so memorably puts it, a good way to think about the metabolic rift is that 'the ecological catastrophe has already occurred'.[24]

Can the metabolic rift be healed? In his 1962 book *The Structure of Scientific Revolutions* Thomas Kuhn introduced the term 'paradigm shift' to describe the ways that scientific worldviews periodically undergo radical change in what appears at the time to be a sudden leap.[25] These 'sudden' paradigm shifts in worldview follow years, sometimes decades, in which scientists have encountered anomalies that don't fit in with the dominant paradigm. Could a paradigm shift in our understanding of 'progress' and 'the economy' be imminent? Are there grounds for optimism that the modernist myth – that the biosphere is a repository of resources to fuel endless growth – will be supplanted by something new?

In the chapters that follow I propose that a new story is indeed emerging. This new story describes an economy based on social energy, using 5 per cent of today's resources, that is not only feasible but will leave the world a better place. This story is not about an imagined future utopia; it's based on actions being taken today that are enabling this new narrative to emerge. According to the German Advisory Council on Global Change (WGBU), the heavyweight scientific body that advises the German Federal Government on 'Earth System Megatrends', a 'global transformation of values' along these lines has already begun.[26] This post-materialist thinking is not limited to rich-world greens. In South Korea, Mexico, Brazil, India, and China, the WGBU found, a significant majority 'supports ambitious climate protection measures' and would 'welcome a new economic system' to achieve that.[27] Although the values described by the WGBU are 'latent' – and numerous laws, and institutional inertia, remain an obstacle – its conclusion is that political and social change on the ground is real, and growing.

This raises an interesting question: if profound paradigm shifts are possible in the worldviews of science; if 'ecological surprises' can transform natural systems, as scientists have shown; and if today's monolithic states could be transformed by 'multi-polarization', as military think tanks predict; in that case, it is surely on the cards that a profound phase shift in *cultural* belief systems is un-paving the way for something entirely new. There's a cheering consequence of this scenario. If, in an age of networks, even the smallest actions can contribute to transformation of the system as a whole, then our passionate but puny efforts so far may not have been in vain. It's like the picture in a jigsaw puzzle that slowly emerges as we add each piece.

2 GROUNDING: FROM HEAL THE SOIL, TO THINK LIKE A FOREST

On a hot day in the foothills of the Cevennes, the mountainous area in France where I live, I'm spreading a mixture of bonemeal, dried blood, crushed oyster shells, and wood-fire ash, onto a growing mound of wood, twigs, leaves, and straw. Each layer is seasoned, as if with salt and pepper, by this powdery mix of minerals and biological activators. The preparation stimulates root growth, soil micro-organism production, and humus formation. Although it takes six of us a day to build one Cevenol mound, our teacher Robert Morez assures us it will supply nutrients to plants, and retain water effectively, for at least four years – maybe more.[1] The invitation had said I would learn 'how to construct a bio-intensive planting mound' – but in my mind I'm making soil, rather than depleting it, for the first time in my life.

During breaks to replenish ourselves, too, with nutrients, I learn that healthy soil is itself a living system – the most dense and diverse medium of interdependent organisms on Earth. There are about 50 billion microbes in one tablespoon of soil; a single shovel can contain more living things than all the human beings ever born.[2] There's a world of connected intelligence down there, too. Mind-bogglingly complex interactions support the flora and food webs upon which we all rely for our existence. In an old-growth forest, millions of super-delicate mycorrhizal fungi are linked together with the roots of plants; these form vast subsoil neurological networks. These interlacing mosaics of mycelium infuse habitats with information-sharing membranes that are aware, react to change, and collectively have the long-term health of the host environment in mind. This vast, invisible web does not just ferry water and nutrients, it also spreads information, and over long distances; a typical mycorrhizal fungal filament can be hundreds or thousands of times the length of a tree root. This chemical communication between plants stimulates their defence against parasites; plants that are not under attack themselves have shown an increased resistance to insects attacking other plants a good distance away. The mycologist Paul Stamets, who describes these networks as 'nature's internet', speculates that fungi may participate in some form of planetary interspecies communication in which we, too, may one day learn to take part.[3] Left to itself, this immense but invisible network is not only self-sustaining, it also determines the metabolic health of all terrestrial ecosystems, including our own. Ninety-nine per cent of all food comes from our soils.[4] As James Merryweather so memorably explains it, all living creatures – animals and plants, bacteria, fungi, and others – are involved in this worldwide, multilayered web of cooperation.[5]

I knew nothing of all this – not a single thing – until that day on the mountain. Remorseful at my own ignorance, but intrigued, I set off to learn more. It turns out that ten thousand years ago, when we discovered that using the plough made farming easier, my predecessors did not realize that intensive tillage fragments these vast but delicate underground networks. They were unaware that fungi and plants depend on each other for survival, and that chopping up the soil disrupts food-producing processes that had evolved over millions of years. In blissful ignorance, we ploughed on regardless – only to discover, over time, that escalating amounts of money, transport, energy, and imported raw materials would soon be needed to feed ourselves. The more food we produced industrially, the greater the damage we caused to soils as a living system. Our use of heavy machinery accelerated the damage; it so compacted soils that plant roots found it ever harder to penetrate – and the soil's capacity to store and conduct water was degraded. As our production increased, ever vaster swathes of land were affected by water and wind erosion. Irrigation with bad water, the escalating use of synthetic fertilizers, and a build-up of salt have made the situation worse. Two hundred years of industrial production have added to the damage in the form of soil contamination: three million major sites around the world have been poisoned by heavy metals and mineral oils. And of course, large areas of healthy land are simply paved over each year as buildings, roads, and airports spread.[6]

The notion that high-tech farming is feeding the world is therefore misleading. The truer story is that industrial agriculture is an extractive industry: it mines the soils for nutrients that are not replaced. We've ruined an area the size of India since the Second World War[7] and, right now, we're losing 3.4 tons of healthy soil a year for every person on the planet.[8] When the first Norwegians came to Goodhue County, Minnesota, the black topsoil was 2 m (7 ft) deep in some places; now, it is only 30–90 cm (1–3 ft) deep.[9] In the UK, scientists have warned that Britain has only 100 harvests left in its farm soil as a result of intense over-farming.[10]

HEALING THE SOIL

What will it take to heal the soil? On its own, soil formation is an extremely slow process – sometimes taking thousands of years – but a growing band of visionaries have discovered that the process can be speeded up dramatically if the right approach is followed. One such pioneer, the Australian soil scientist Dr Christine Jones, has demonstrated that new topsoil will form rapidly, and naturally, with the right combination of biomass and turnover of plant roots. In what she calls her 'Rules of the Kitchen', Jones lists six essential ingredients for soil formation: minerals; air; water; living things in the soil – such as plants and animals, and their by-products; living things on the soil, ditto; and what she describes as 'intermittent and patchy disturbance regimes'. 'In order for new soil to form, it must be living,' Jones explains; 'life in the soil provides the structure for more life, and the formation of more soil. That's why healthy groundcover, high root biomass, and high levels of associated microbial activity, are fundamental to building new topsoil.' Farmers using cover crops as green manure can produce 1 cm (½ in.) of topsoil in three to four years. Even better: when the value of the crops used in this approach is factored in, the net cost of restoring soil is negative.[11]

These principles have been shown to work on a large scale in a project in Zimbabwe called Operation Hope.[12] More than 2,600 hectares (6,500 acres) of parched and degraded grasslands have been transformed into lush pastures replete with ponds and flowing streams – even during periods of drought. Surprisingly, this was accomplished through a dramatic *increase* in the number of herd animals on the land. Behind Operation Hope is an approach called holistic management, applied to rangeland practice, that has been developed over fifty years by Allan Savory, a former wildlife biologist, farmer, and politician. Savory's method is based on a singular insight: grasses can't graze themselves. Before man came along, herbivores co-evolved with perennial grasses. When a large herd moved around freely – accompanied, that is, only by pack-hunting predators – they dunged and urinated with very high concentration on the grass. No animals like to feed on their own faeces, so they had to move off of their own faeces within one to three days and they could not return until the dung had weathered and was clean again.

Moving across the land in large herds, the herbivores trample and compact soils while also fertilizing the soil with concentrated levels of nutrient-rich animal wastes. This approach aligns itself with nature in a comprehensive way; it increases plant growth and also re-establishes livelihoods through additional livestock, while increasing wildlife populations through holistic management. Grasses depend on herbivores to help them with their decay process. When large herbivores such as kudu and Cape buffalo disappear, grasses begin to decay far more slowly through oxidation. When millions of tons of vegetation are left standing, dying upright, light cannot reach growth buds; the next year, the entire plant dies. The death of grass leads to bare ground, and the desert spreads.

Savory was not alone in understanding the importance of compacting on the health of vegetation. In the early 1970s, agricultural institutes in Texas and Arizona designed machines to simulate the physical effects of once prevalent vast herbivore herds such as the millions of bison that roamed North America. Machines with names like the Dixon Imprinter were used on thousands of acres of the western US to break soil crusts and cause indentations and irregularities, while laying down plant material as soil-covering litter vital to soil health. Imprinting, as the technique is called, is still practised; agriculture labs in various countries have developed rollers that imitate the hoofprints of passing buffalo and trample green manures and old stalks into the ground.[13] Trouble is, these approaches do not heal the soil. Their machines are too heavy. Although a big buffalo weighed about a metric ton (2,200 lbs), the monster tractors used in mega-agriculture can weigh 45 metric tons (over 100,000 lbs). Machines this large do terrible damage to the soil underground in a single pass.

For Allan Savory, the hooves, mouths, and digestive systems of real animals do this same task more effectively. The process consumes no fossil fuels, and can be repeated continuously at no cost. Large herbivores break soil crusts, but without damaging the subsoil, and the broken crust allows soil to absorb water and to breathe; this enables more plants to germinate and establish. The effect is more pronounced when animals are concentrated in large herds – which is how they behave when under threat from pack-hunting predators. Operation Hope therefore runs livestock in what Savory calls a 'predator-friendly manner. We don't kill the lions, leopards, hyenas, wild dogs or cheetah because their presence is crucial to keeping wildlife moving and thus the land healthy.' Livestock are held every night in portable lion-proof corrals (known as kraals in southern Africa). Large animals also compact the soil under their hooves – 'anyone who has had a horse stand on their boot understands this', jests Savory – but it's the right amount of compaction for good seed-to soil-contact, which increases germination. The need for compaction is why gardeners tamp down the soil around seedlings or seeds.

Ruminants also return standing grass-plant material to the soil surface earlier than the same plant material would have returned to the soil had the animals not been there. One has only to watch a cow or buffalo trample or dung to know this. In short, the conversion of plant material to litter or dung is essential to maintain biological decay. Machines designed to imitate animals cannot do this.

TIME, NOT NUMBERS

Grasslands where rainfall is seasonal require periodic disturbance for overall health – but not too much, and not too little. Overgrazing is a function of time, not of animal numbers alone. Trampling for too long turns the soil into powder, which increases erosion by wind and water; and dung and urine, like most things in excess, become pollutants when animals are there too long – a lesson industrial-scale 'feedlot' cattle farmers soon learn. Whether there is one cow or a thousand, Savory explains, is not so important; the important variable is *time*. Moments of high physical impact – trampling, dunging, and urinating – are choreographed in short periods between much longer periods for plants and soil life to recover. As a guide, three or fewer days of grazing are followed by three to nine months of recovery – but, because they manage holistically, Operation Hope's herders do not follow abstract time regimes. Each piece of land, and each moment in time, is unique.

Savory's use of increased livestock to reverse desertification is a profound challenge to mainstream approaches to land use and agricultural development. For although the Green Revolution increased global food production tremendously, its reliance on fertilizers, intense watering, and heavy machinery degraded its ecological base, and its associated social systems, in the process. In the pursuit of efficiency and increased output, so-called production agriculture relied on massive inputs of petrochemicals and herbicides, focused on just one crop at a time, and confined large numbers of animal into grim 'feedlots'. The good news, according to Savory, is that this damage can all be reversed by what he calls a 'Brown Revolution' based on the regeneration of covered, organically rich, biologically thriving soil, and brought to fruition via millions of human beings returning to the land and the production of food. 'Viewed holistically biodiversity loss, desertification, and climate change, are not three issues, they are one,' Savory says. 'Without reversing desertification, climate change cannot adequately be addressed.' The more humid and biologically productive regions of the world need to develop agricultural models based on small, biodiverse farms that imitate the natural, multi-tiered vegetation structures of those environments. This is where most of tomorrow's grain, fruits, nuts, and vegetables will be produced, as well as most of the dairy products, and some of the meat. Savory's approach has big social benefits, too. Globally, small-scale livestock production employs 1.3 billion people and sustains livelihoods for about 900 million of the world's poorest people – many of them women. They will have a vital role to play in the restoration of degraded soils.

Although Savory describes these insights as common sense, he has spent fifty years battling to make the scientific case for his approach. For most of his life, he has had to contend with intense opposition from agricultural researchers intent on 'proving' it does not work. Savory's belated acceptance by the mainstream is one sign of a profound shift in scientific understanding of energy and nutrient transfers in ecosystem ecology. What Savory learned on the range is confirmed by biological studies of plants, animals, terrestrial, aquatic, and marine ecosystems and how they interact with each other. Systems can have properties as a whole, it turns out, that are not explicable in terms of the sum of parts that scientists once studied in isolation. The drive to scale up food production was a powerful incentive to bypass complexity, but a management approach that works well in car factories or software has turned out to be self-defeating when applied to the land.

THINKING LIKE A FOREST

If maintaining the fertility of the soil is a core principle of ecological agriculture, so, too, is a commitment to think in longer timeframes than markets – or even than individual human lifespans. We need to think less like a machine and more like a forest. At Windhorse Farm in Nova Scotia, James W. Drescher is the latest custodian of an experiment called 'enrichment forestry' that has been in progress for four generations – just a blink of the eye in the life of a forest. 'Windhorse is on the leading edge of something very old,' says Drescher; 'wealth, from the forest's point of view, is biological material.' Because a healthy forest is rich in biodiversity and heavy with stored carbon, the key to its long-term health is the retention of wealth after it has been created. Conserving that wealth, Drescher has learned, is dependent on the very slow decomposition of huge volumes of dead wood. Dead wood is the life of the forest, Drescher explains; almost half the animals in an old-growth forest live in or on or from it. Foresters who act as land stewards, rather than like factory managers, are therefore selective in deciding which trees to harvest and remove. Most dead trees, or trees that have fallen naturally, are left where they are. By harvesting only the slowest-growing trees in a stand, the forest's overall vitality is increased. In a similar spirit, the tallest trees are never cut; this increases canopy height. Species of tree that are under-represented in a particular stand are left alone to conserve species diversity. Pathways in the forest are lined with sawdust and bark, not with concrete; animals and plants travel and disperse along these corridors of connectivity. Remarkably, this 'forest health first' approach is economically viable – more so, in fact, than the clear-cutting approach of mainstream commercial forestry. If a 40-hectare (100-acre) lot in the Acadian Forest had been clear-cut in 1840, and again in 1890, 1940, and 1990, Drescher explains, the total harvest would have been much lower than the wood harvested by the annual selection methods; and, of course, there would be no standing merchantable timber at all today.

In today's culture of short-term profits, the wisdom and skills needed to maximize the yield from a forest over a period of a century or more are rare. But looking forwards, the Windhorse Farm experiment is proof that it's possible to make a living in ways that respect, and not harm, other life forms that are also trying to make a living there. The forest itself – not the timber that's sold – is the primary product. In that sense Windhorse forestry is a set of principles rather than a model to be replicated at will. It's a practice that demands diligent study, keen observation, insightful analysis, and resourceful generosity. Drescher describes as 'deep stillness' the everyday practice in which foresters, woodlot owners, and other workers simply hang out in the forest a lot more: studying, observing, reflecting, working, and, as Drescher puts it, 'investing lots of time doing as close to nothing as possible'.[14]

If holistic rangeland management and do-little forestry sound fringe – well, they are, for now. But in a growing number of real-world contexts, the respectful interdependence of people and living systems is coming back to life. I'll tell you about more examples later in the book, but my purpose here is to suggest that reconnecting with the land and proactive soil restoration are set to become mainstream. At the Stockholm Resilience Centre in Sweden, Per Olsson and his colleagues are amassing a growing number of stories in which groups of interested parties inhabit their land in healthy ways.[15] Olsson describes these examples as 'social-ecological systems' in which often diverse communities are finding ways to share rights, responsibilities, and power in ways that put the interests of the land and its soils first.

BIOREGIONS

What researchers describe opaquely as 'adaptive ecosystem-based management' is at heart a social and cultural process, not a technical one. A sense of belonging, and shared responsibility for the land, is the social glue that binds diverse groups together. A new political and geographical concept – the bioregion – is beginning to strengthen these shared ties.

National boundaries are an outdated way to inhabit the land. In the Global South, where lines were often drawn literally in the sand by former colonial powers, the relationship between city and countryside is especially distorted; nearly half of the 250 biggest cities in the Global South were founded by European colonial administrations. In the United States, too, most state and county boundaries were drawn as straight lines on a map by people who did not know the land. What's emerging now is an approach to the governance of cities and their region, based on place, that enables the regeneration of soils, watersheds, and biodiversity. A bioregion is literally and etymologically a 'life-place', in Robert Thayer's words, that is definable by natural rather than political or economic boundaries. Its geographic, climatic, hydrological, and ecological qualities – its metabolism – are complex, and unique.[16] A bioregional approach reimagines the man-made world as being one element among a complex of interacting, codependent ecologies: energy, water, food, production, information. It attends to flows, biocorridors, and interactions. It thinks about metabolic cycles and the 'capillarity' of the metropolis wherein rivers and biocorridors are given pride of place.[17]

A growing worldwide movement is looking at cities through this fresh lens – but the lens is not a rose-tinted one. Modern bioregionalism does not seek a return to pristine nature or an unspoiled 'before' – as if ecological change could be reversed. The sense, instead, is our wellbeing is intimately connected to the vitality of living systems; we should make them – and the interactions between them – the focus of our efforts. Bioregions are not a form of wildlife park; they embrace the urban landscape itself as an ecology with the potential to support us.[18]

STEWARDSHIP

Changes in policy are responding to a powerful cultural shift in which the concept of land use is giving way to land *stewardship*. The word 'steward' (from the Anglo-Saxon *stigweard*) originally meant 'keeping in trust for the absent king'. It evolved to include managing an estate on behalf of an absent owner, but is most commonly used today in connection with stewardship of the environment. Organizations such as the Forest Stewardship Council promote responsible management of forests; the Marine Stewardship Council promotes sustainable fisheries and the long-term interests of people who depend on fishing for livelihood or food; the Countryside Stewardship Scheme in England sustains landscape beauty and diversity, protects and extends wildlife habitats, conserves historic features, restores neglected land,

and improves opportunities for people to enjoy the countryside. A growing number of educational projects link nature and culture and promote learning about the intimate linkages between them. A recent EU programme called LandLife, for example, promoted land stewardship as a means for national governments to meet to agreed biodiversity goals.[19] And in Turkey, hundreds of teachers across the country are being certified as ecoliteracy instructors in a programme that spans subjects from soil erosion to ethical forestry. Their classroom is an arboretum.[20]

In mainstream land management, stewardship is creeping in literally from the edge with the development of habitat networks in productively marginal areas. Edge habitats such as hedges, ditches, and banks, waterways, abandoned fields, and forest sites, are all havens for biodiversity; they provide forage plants at the start and end of the nesting season when flower-rich grassland areas are otherwise being grazed or have been cut. Edge habitats are also nesting and hibernation sites; they provide relatively sheltered and undisturbed conditions with plenty of tussocky areas and abandoned rodent holes. They also play a vital role in connecting up larger areas of habitat in the landscape. In the UK, an organization called Hedgelink involves farmers, planners, environmentalists, and local communities in a nationwide Hedgerow Biodiversity Action Plan. Volunteer groups collect data on a wide variety of variables – from the age of a hedgerow, or the presence of ditches, to the types of soil in a hedge or the location of gaps.[21] Other landscape niches for biodiversity include schoolyards, sacred groves, parks, areas around roadways, industrial and hospital sites. The Danish government is promoting the expansion of natural 'field margin ecotones' – buffer zones between highly cultivated fields that are kept pesticide and additive free. Just how wide these buffer zones should be is of course a contested issue, but one government report advocated a minimum width of 6 m (20 ft). Such zones also increase the supply of food for game birds and hence enable extra income for landowners.[22]

In Scotland, where a Centre for Stewardship has been established on the Falkland Estate, Ninian Stewart is convinced that the time is ripe for a new model of stewardship that, in his words, 'draws from the past and seizes our day to leave a sustainable legacy for the future'. Stewart's approach widens what he calls the 'circle of consideration' further into the future and away from self-interest than is typical in today's stewardship regimes. We need, he says, to 'restrain our present-day kings from headlong exploitation, depletion and destruction of our social and biological capital'.[23] 'The world is calling out for more responsible long-term thinking,' Stewart told me when we corresponded; 'in an age when speed, profit-taking and consumption are undermining the sustainability of the world as we know it, we would be wise to adopt more of the mindfulness, long-term ethical investment and care for the wider community that are the hallmarks of stewardship.'

A bioregion makes sense at many levels: practical, cultural, and ecological. By putting the health of the land, and the people who live on it, at the centre of the story, a bioregion frames the next economy, not the dying one we have now. Because its core value is stewardship, not perpetual growth, a bioregion turns the global system on its head. Rather than drive the land endlessly to yield more food or fibre per acre, production is determined by the health and carrying capacity of the land through time – a factor which is constantly monitored. Decisions are made by the people who work the land, and know it best. Prices are based on yields the land can bear, and on revenues that assure security to the farmer. 'Growth' is measured in terms of land, soil, and water getting healthier, and communities more resilient.

The idea of a bioregion begins to heal the metabolic rift I wrote about earlier. It reminds us that the cities we live in now do not exist separately from the land they are built on. The idea of a bioregion is tremendously motivating, too, in ways that abstract words like 'sustainable' are not. The word triggers people to seek practical ways to reconnect with the soils, trees, animals, landscapes, energy systems, water and energy sources on which all life depends.

The management of bioregions and 'whole landscapes' is complex, of course. A bioregion cannot be divided neatly into the planning categories of a city: Centre, Periphery, Rural; Work, Rest,

Play. Bioregions are a mosaic of both natural and human-modified ecosystems that change constantly as ecological, historical, economic, and cultural processes interact.[24] Their size can vary enormously, too – from hundreds to tens of thousands of square kilometres. No rule books exist for the governance of a bioregion: each community has to write its own.[25]

The tools for bioregional governance are in development. Colleges across the north-western United States have developed a Curriculum for the Bioregion that transforms the ways in which tomorrow's professionals will approach place-based development. The curriculum, which is taught by experts from across the Puget Sound and Cascadia bioregions, is divided into such topics as Ecosystem Health; Water and Watersheds; Sense of Place; Biodiversity; Food Systems and Agriculture; Ethics and Values; Cultures and Religions; Cycles and Systems; and Civic Engagement.[26] A treasure trove of completed projects is further evidence that these are not just academic subjects. Multidisciplinary teams have evaluated water-quality data as indicators of the health of an ecosystem; mapped stream channels in a local watershed; learned about the geology, hydrology, soils, and slope stability of a local town; analysed the environmental costs of metal mining; studied how indigenous peoples used to inhabit their region – and discussed how best to integrate this wisdom into new models of development. The thinking behind bioregions, if not yet the name, is now shaping top-down policy, too. More than fifty governments and major institutions – from the African Wildlife Foundation, to the World Bank – have committed to pursue a so-called 'whole landscape approach' in their approaches to sustainable development.[27]

SOIL AND SOUL

Thinking and acting at the scale of a bioregion has a spiritual as well as a practical dimension. We are born with an inherited aesthetic tendency to appreciate an intimate connection with the world. But you may be wondering, as you read this, how soil health can possibly be interesting to modern people, especially if they live in cities; most urban people think far more about connecting with each other than about connecting with the soil. In a world where fewer than half of us ever see or touch the stuff, asking city dwellers to empathize with earthworms sounds like too big an ask.

For many years I harboured the same misgivings – but then I had an epiphany on an island in Sweden. Fifty designers, artists, and architects, gathered together for a summer school, were asked to explore two questions: 'What does this food system taste like?' and 'How does this forest think?' My concern that living soil would not engage these city-based designers proved unfounded. It was like pushing at an open door: our students went scrabbling around the forest of Grinda like so many voles. They found ways to catch the taste of the forest and put it in a pot. They made cookies with forest berries and bartered these with tourists. They created tactile pathways so we could feel the forest through our feet. A Latvian designer made pine-cone syrup and gave it to Teacher, who was mightily pleased. One team invented a Soil Tasting Ceremony. They made infusions from ten different berries on the island and displayed them next to soil samples taken from each plant's location; the soils were displayed in wine glasses. We were then invited to compare the tastes of the teas and soils in silence. It was a powerful moment. Systems thinking, I concluded, becomes truly transformational when combined with systems *feeling* – which is something we all crave. 'We yearn for connection with one another, and with the soul,' writes Alastair McIntosh, 'but we forget that, like the earthworm, we too are an organism of the soil. We too need grounding.'[28]

3 WATERKEEPING: FROM HARVEST THE RAIN, TO RIVER RECOVERY

In Brazil, the monumental jequitibá tree moves hundreds of gallons of water up into its canopy every day. It does so without pumps, or electricity, or any of the costly and resource-intensive infrastructures upon which modern cities depend. The jequitibá is a living emblem, for me, of an emerging future in which the ways we use and think about water are once again shaped by knowledge of place and watershed, and not by the power-hungry mechanics of fetching it from somewhere else. For centuries, we've dug, dammed, channelled, and concreted over rivers and watersheds without much thought for the consequences. We've built cities on top of them. We've dumped pollutants into them. We've pumped up ancient groundwater a million times faster than nature can replenish it. And we've built dams – 48,000 large ones, and counting[1] – that displace whole populations and disrupt the hydrogeological balance of watersheds, estuaries, and lagoons. Our extractive use of water on an industrial scale has accelerated climate change, too. When water is moved from where nature put it, in watersheds and aquifers, rivers no longer reach the ocean, aquifers run dry, and deserts expand. And when vegetation is removed from the land, so, too, is the green biomass that once absorbed the rain; cloud vapours blow away and deserts replace living ecosystems.

The jequitibá would be an inspiration to the world's designers and city builders, faced with a growing precarity of water supply, if only we were minded to notice. But we're not. Thanks to the metabolic rift I described in Chapter 1, we've lost touch with the reality that water is a living system; if we think about water at all, it's as a liquid that comes from a tap. As a measure of our divorce from the lived reality of water, try this thought experiment. Imagine emptying 500 1-litre (34-fl oz) bottles of water into a huge pot and carrying it 50 miles (80 km) – every day of the year. Does moving all that water around sound hard? It *is* hard – but that's how much water is moved every day for a US citizen, and how far, once her share of the agriculture, manufacturing, car washes, window cleaning, laundries, ornamental ponds, health clubs, swimming pools, and golf courses are added together. A vacationer at a tropical resort uses as much water in one day as local people do in 100. It takes 2,700 litres (700 US gallons) of water to grow the cotton used in my T-shirt; 10 litres (2½ gallons) to manufacture one sheet of office paper;[2] 1,500 litres (400 gallons) to grow enough biofuels to move one car 10 km (6 miles); 140 litres (37 gallons) to grow enough beans for a cup of coffee; 1,455 litres (384 gallons) to make a pizza margherita;[3] and 4 litres (1 gallon) to produce *one single almond* in California's Central Valley – and, as I'll explain in Chapter 5, 85 per cent of those almonds are exported to China and India.

Capitalism is not uniquely to blame for this madness. Pre-industrial societies were no more inclined than we are to leave water in its natural state. History is replete with mind-boggling tunnels, aqueducts, and dams.[4] In Egypt, *qanats* linking chains of wells irrigated vast areas of fertile land west of the Nile – and that was in 500 BC; in Iran today, 20,000 qanats are still in operation 3,000 years after they were first developed. During his reign as king of Sri Lanka from 1153 to 1186, Parakramabahu asserted that 'not even a little water that comes from the rain must flow into the ocean without being made useful to man'; the busy king went on to construct or restore 65 dams, 3,910 canals, 163 major reservoirs, and 2,376 minor tanks – all in a reign of 33 years. Parakramabahu started a tradition whereby every Sri

Lankan king would build dams; the island now contains more than a thousand. No country in the world contains so much man-made irrigation per square kilometre.[5]

In cities, especially, we've been disconnected from water as a living system since Roman times – and by design. That was when we first figured out how to 'get rid' of water using engineered structures to do so. The Romans were the first to design rapid-transit water conveyance systems, integrated into the built environment, that kept land relatively dry, provided a supply of drinkable water, and carried away human waste for disposal. Over centuries, these hard solutions – sewers, drains, treatment plants, and the like – have steadily disappeared from sight under growing areas of what ecologists today call Impervious Surface Area (ISA) – all those roads, car parks, airports, buildings, driveways, and sidewalks. China has more ISA in total than any other country, but US citizens have the highest amount per person: 297 square m (3,200 square ft).[6] This total matters less than its location: most ISA occurs in primary drainage basins where its greatest man-made impact on watersheds has occurred. In India, they say that rivers are the birthplace of civilization – and civilization the graveyard for rivers.

All this infrastructure remained out of sight and out of mind for centuries until we realized, belatedly, that, with climate change, a lot of our water infrastructure is no longer fit for purpose.[7] As rainfall becomes more intense but less frequent, for example, existing reservoirs are proving too small to hold the extra water, and downstream flooding is more likely to occur as a result; then wet and angry house owners scream at elected officials that 'something must be done' to protect their biggest asset – even though that 'something', in their mind, is the kind of hard structure that we can no longer afford. Besides, after a thousand years of civil and water engineering efforts to obliterate them, we are now discovering that ponds and vegetation, natural stream courses, buffers, and floodplains are as a sign of urban health.

SUDS IN THE CITY

It sounds like a huge change – but a profound transition from hard-engineered systems is fast gathering pace. The new paradigm in water management – so-called Water Sensitive Urban Design – features a return to the hydrology of a city as it was before the concrete conveyor system was built. The focus has shifted from high-entropy engineered solutions, such as reservoirs and sewer networks, to softer ecological systems that give priority to water where it falls; this small and local approach conserves water, improves water quality, reduces flooding and erosion, and promotes revegetation. At street level, Sustainable Urban Drainage Systems (SUDS) involve the redesign of roofs, pavements, streets, and parking spaces. In their place come rain gardens, surface wetlands, restored ponds and streams, reed beds, and worm colonies. Some new equipment is needed to make cities 'water-wise', of course: rainwater tanks; large bladder storage systems; greywater plumbing; settling tanks, physical filters; smart sensors for systems monitoring; new systems for maintenance. But equipment is a relatively small part of the solution; water-harvesting practices are for the most part simple, low-tech activities.

Physical work and grassroots social organization are the most important ingredients. With budgets at local and national level under extreme pressure, a growing number of communities are taking hands-on measures to restore urban watersheds house by house, and street by street. In Tucson, Arizona, which has only 305 mm (12 in.) of rainfall per year, residents are becoming active participants in water conservation and water-harvesting together with non-profit organizations and local businesses. One pioneering non-profit, the Watershed Management Group (WMG), started as a tiny seedling of an idea in the minds of five students graduating from the University of Arizona's Watershed Management programme.[8] They noticed that much of the focus of environmental programmes was directed at rural areas, protected parks, and wilderness areas – but not at the city streets where they all lived. The urban landscape became their focus. Through its co-op programme, WMG helps homeowners harvest rainwater on their own properties at minimal cost. The way it works is that a homeowner joins the co-op, volunteers time on other water-harvesting projects, and accrues a set number of hours. When enough hours have accumulated, the volunteer gets to host a workshop at his or her own house and reap the benefits of the team's labour. Between six and fifteen people are involved in a typical workshop, including at least one expert staff member. WMG now has twelve demonstration sites across Arizona where citizens can see water-harvesting practices in real-life contexts. The group has also developed its own training scheme: its Water Harvesting Design Certificate covers green infrastructure, advanced cistern applications, advanced greywater applications, and small-scale erosion control and riverbank restoration. With its emphasis on integrated and sustainable design, a growing number of architects, landscapers, planners, entrepreneurs, and community organizers have acquired this know-how to retrofit residential and commercial sites.

The power of grassroots initiatives is making itself felt in India too. For S. Vismanath, a thirty-year veteran of water stewardship projects, 'hyper-local leadership' is the key to solving the sanitation and water challenge in the world's megacities – even when the battle seems already to have been lost. In Bangalore, for example, once known as the city of a thousand lakes, the ravages of urbanization, sewage dumping, and encroachment have left barely thirty-four in their original healthy state.[9] But, Vismanath told me, a growing number of success stories is evidence that the tide may, literally, be turning. An especially inspiring movement called The Ugly Indian, started by design students in Bangalore, is about cleaning up the informal latrines, trash-covered sidewalks, and illegal rubbish dumps that pollute so much of the water supply.[10] Small groups take unilateral action to clean up a stretch of sidewalk, for example, and then they talk to everyone involved in that street: the garbage collectors, shop-owners, municipal cleaning staff, office workers who dump trash on the street, and so on. They use the visible result of their initial action to start conversations – 'Look how clean our street could be!' – and ask people who are part of the problem (which is pretty much everyone) to imagine themselves as co-owners of a clean street, not a filthy one. Ugly Indians don't blame their fellow citizens, or politicians, or 'the system'. They act first, and then they talk. They make it 'our' problem, not 'your' problem. This combination of social skills with systems thinking is remarkable. The Ugly Indian movement has spread to a dozen Indian cities, and a similar project has started in Karachi.

RIVER RECOVERY

The urge to reconnect with rivers and watersheds resonates in the most unlikely places – none more so than Mexico City. A young architect friend of mine, Elias Cattan, is consumed by the idea that reconnecting with its rivers is the best way for his city to move forward.[11]

Like Bangalore, Mexico City was once a city of interconnected lakes and more than sixty rivers – but is now a dusty megalopolis covered by busy roads and shopping malls. The Aztecs were the first to alter the natural hydrological rhythms of the city – partly to manage flooding, partly to create a capital hub that could be easily protected – but industrialization, and the rise of the car, amplified that initial damage as urban planners treated rivers as a problem to be overcome by hard engineering; waterways were forced into channels, covered, and built over. Now, together with a team of ecologists and architects, Cattan's design firm has submitted a proposal to replace busy roads with a ring of water and parks around the city centre, restoring at least three rivers in the process; these would be home to myriad plants, fish, and birds.[12] City officials love the idea, but say they lack the resources to implement it. Elias thinks they are too pessimistic, and tells them about projects of equal ambition that have been implemented elsewhere. Local officials were especially impressed, Cattan tells me, by the example of Seoul, in Korea; that city's mayor, Lee Myung-bak, removed a major freeway to make way for the Cheonggyecheon River.[13]

In London, ambitious plans to restore hundreds of miles of hidden rivers are driven by the need to adapt to climate change: to create floodplains that will protect local homes and businesses in the event of heavy rains.[14] As with Bangalore and Mexico City, London has a lot of catching up to do. Although the Domesday Book of 1086 recorded more than six thousand mills and freshwater fisheries on London's rivers, streams, and brooks, London has been burying rivers since medieval times – at first, because so many became dumping grounds for rubbish and sewage. Many of London's rivers and waterways were paved over after the 'Great Stink' of 1858, when the whole city was choked by the smell of the fetid Thames. More civil engineering accretions followed during the twentieth century to combat flooding and facilitate urban development. The result, over the centuries, is that more than 400 km (250 miles) of once natural rivers have been artificially modified in one way or another. Dave Webb, an ecologist who coordinates river restoration efforts in London, needs an amazing array of skills to do so. 'When we survey a river, we use a technique called geomorphology to map the channel's shape. We study its ecology and habitats. We measure the diversity and biomass of invertebrate, and the richness (or otherwise) of marginal plant community.' But ecologists don't just study life forms in swamps: as Webb explained to Juliette Jowit in *The Guardian*, people skills are also important. Webb has to deal with business, industry, community groups, politicians, community groups, volunteers, and fellow scientists. 'It's a bit like herding cats,' he says, 'only in water!'[15]

GREEN LIVERS

Plants, as well as manual work, are proving to be an important new resource in these emerging approaches to water infrastructure. In China, for example, plants are being used to clean up watersheds polluted by China's heavy industry. Professor Peter Saunders describes clusters of such plants as a waterway's 'green liver'.[16] Many of the pesticides, solvents, dyes, and other by-products emitted by agriculture and industry are eventually transported to natural vegetation and cultivated crops; if people eat the plants, these toxins can accumulate in their organs with dire effects on health. But plants can also metabolize harmful substances, in a process called rhizofiltration, and in China alone more than 400 species have been identified as having potential for soil and water remediation. The process requires no external inputs; once a wetland phytoremediation system is in place, plant photosynthesis is naturally self-sustaining. In Kunming, in south-west China, water hyacinths are the favoured plant for this task. In their voracious quest for nutrients, water hyacinths absorb a good amount of the nitrogen and phosphorus found in domestic sewage; these wastes accumulate in the plant's roots, which then become valuable compost or organic fertilizer. Because water hyacinths thrive on sewage, they have exciting promise as a natural water-purification system at a fraction of the cost of a conventional sewage treatment facility.

In Uzbekistan, the favoured plant for water treatment is liquorice.[17] In the country's north-west, hundreds of farming communities were forced in recent times to abandon 30,000 hectares (74,000 acres) of land degraded by over-irrigation in the Aral Sea Basin. (As you will read in Chapter 6, this ecocidal practice was carried out to produce the cotton in my two-wash-two-wear T-shirt.) The same farmers have now started to cultivate wild liquorice on salt-ridden soils near drainage canals. The tops of the shrubs, established either from roots or seed, are cut for livestock fodder; by the third year, some farmers dig up the roots to export for profit to Japan, South Korea, and the Ukraine. Extract from liquorice roots is used in medicines, candy, food, alcohol, and even cosmetics. As explained by soil scientist Andrew Noble, the key is the deep-rooted nature of liquorice; its ability to lower the water table prevents salt from rising to the surface of the soil. In effect, salt collected in the soil is gradually flushed out, enabling the land to be irrigated again for new crops.

As in China and Uzbekistan, so, too, in Europe. I was startled, during a visit to a civil engineering trade show, to come across a stand called British Flora amid displays, in the rest of the show, of massive pipes and pumps.[18] In place of large-scale concrete, I learned, British Flora specializes in bio- and phytoremediation, watercourse and bio-engineering, wildlife, and saltmarsh and maritime habitat creation. Its clients, remarkably, are civil engineers, hydrologists, environmental consultants, landscape contractors, planners, and government bodies. Fundamental to British Flora's success is its expertise in species selection, genetic purity, and crop production. Its designers understand the behavioural properties of native plants within different habitats, and talk animatedly about which native plants and seeds will tolerate high salinity levels, or best absorb heavy metals or hydrocarbons.

ONE MILLION CISTERNS

In Brazil, where millions of poor farmers struggle to survive on semi-arid land, a respectful interdependence between people and living systems is also coming to life. In arid and semi-arid climates, where solar radiation is intense and evaporation rates are high, storage in surface reservoirs can result in large-scale water loss. In a remarkable programme called Articulação no Semiárido Brasileiro (ASA), 22 million people – in a region as big as Texas and California combined – are being helped to harvest the 750 mm (29½ in.) of rain that fall on this vast semi-arid region each year.[19] The idea is to store water for the nearly eight months of the dry season, make optimum use of the limited supply available, and prevent evaporation. Hundreds of community and religious institutions, cooperatives, churches, NGOs, and rural worker federations have helped local residents build 350,000 household water tanks and 9,000 irrigation tanks since 1999; the target is one million cisterns by 2020.[20] ASA is another example of a social and ecological approach in which artefacts are constructed by a process designed to help communities adapt to challenging climate conditions, while respecting the bioregion's ecosystems and traditional cultures. The way the project is governed is as important as the design of the cisterns: groups of installation projects are implemented by 'micro-regional management units', which bring together several municipalities selected by public process.

The convergence of social and ecological thinking is evident in another inspiring example, in Hawaii, which I learned about from the ecological designer Howard Silverman.[21] In Hawaii, where populations of rare tropical fish were threatened by overfishing by the aquarium industry, a 'lost fish coalition' was able to designate over 30 per cent of coastal waters as off limits to aquarium collection.

The success of this against-the-odds agreement was attributed by Silverman to the traditional *ahupua'a* system of land division and community responsibility – an ancient custom that has proved far more appropriate for modern times than anyone would have imagined in advance. It seems to work because resources are divided into smaller areas among local people who know the region and have a stake in working together. This kind of integrated approach, which improves the governance of land, water, forests, and grasslands, is called 'Multi-Actor Ecosystem Participation'.[22]

HANDS-ON WATER STEWARDSHIP

In drought-affected parts of Andra Pradesh, in India, a proliferation of illegal boreholes threatened the long-term viability of aquifers across the whole region. When regulations imposed from afar proved ineffective, smallholder farmers in more than 600 villages learned how to manage local watersheds as a common resource. Some 20,000 farmers now double as barefoot water technicians in a social practice described by resilience researchers as Participatory Groundwater Management (PGM).[23] The PGM approach is based on small 'hydrological units' – a unit being a cluster of micro-watersheds. Hydrological data is collected for each hydrological unit – daily rainfall, water levels, borewell discharge, and daily stream flows. Being on the spot, smallholders are well placed to use this data to improve their understanding of how well groundwater resources are recharging at a local scale. In addition to on-the-spot measurements, farmers have also started to use GPS to collect data remotely from wells, rain gauge stations, and artificial groundwater recharge structures. The information is shared communally; rain gauge stations display the amount of rainfall received on a day-to-day basis. These display boards act as village discussion points on issues such as drought, floods, water levels, and agriculture practices. Trusted and independent information is a key to the success of PGM. Before the new system was introduced, farmers had to rely on data provided by so-called 'input dealers' – fertilizer, seed, and pesticide companies; these sources tended to downplay the huge wealth of grounded knowledge.

Community-based institutions occupy a central position in this smallholder-centred approach. In Farmer Water Schools, for example, farmers and scientists explore different management options on an equal basis. Data analysis is continuous, multi-dimensional, and multi-scalar. For each hydrological unit a so-called 'base document' serves as a reference point for farmers and support agencies as they plan future activities. Water resources are described at an ecosystem level: climate, rainfall, drainage, groundwater, land use, and so on. The document also records soil conservation activities, the range of crops being used, which livestock are being reared, their yields and net returns, and so on.

BIOREGIONAL SCALE

In Germany, a project called Aqualon is a developed-world model for social-ecological water management at a regional scale.[24] Covering a 200-square-km (77-square-mile) area surrounding the Great Dhünn Dam – the second-largest drinking water reservoir in Germany – Aqualon brings together thirty participating municipalities in partnership with researchers from the region's universities. Among the latter is a team of graduate landscape and architecture students from TU Graz and the University of Wuppertal, led by Professor Klaus K. Loenhart. Their contribution to Aquilon has been to develop a series of social-ecological scenarios for the watershed as a whole. These include novel forms of agriculture, sustainable forestry, land-based municipal sewage treatment, rainwater harvesting – even the settlement of beavers as ecologically valuable animals. The TU Graz project has two phases. The first is an investigation and mapping of the region's biodiversity, cultural practices, and patterns of local production; this creates an initial schematic of the various economic flows and ecosystems. The second phase involves looking for ways in which these different resources could be developed, and complement each other, in a future metropolitan regional system. The guiding principle here is that natural processes and human activity, connected by enabling technologies, will interact in ways that regenerate and reconnect, rather than extract and degrade.

In Sweden, these connections between social and ecological processes are not being left to chance. At the Stockholm Resilience Centre, researchers led by Arvid Bergsten compare maps of natural systems with administrative maps – and look for gaps. In one project, Bergsten's team mapped the level of ecological connectivity between 641 wetlands in municipalities neighbouring the city, and then compared this to patterns of day-to-day communication among the city's 26 municipalities. They found that bits of government responsible for land use planning were not communicating well with agencies looking after wetlands. The mapping exercise helped to improve coordination between the different city government agencies whose actions impact collectively on watershed health.[25]

WATER AS A SOCIAL AND ECOLOGICAL SYSTEM

The examples I have described above are proof, for me, that a latent passion to reconnect with rivers and watersheds can unleash tremendous social, cultural, and design energy. These projects do not portend a U-turn back to premodern ways; new technologies and materials are an important part of the picture. But the main difference between this new course and the hard approach to infrastructure it replaces is that the health of living systems is the main inspiration behind the actions these people are taking.

This interaction between the social and the ecological is not a novelty; it has deep cultural roots that we can learn from. Balinese farmers have been growing rice on terraces since at least the eleventh century. Because the island's volcanic rock is rich in mineral nutrients, water running off mountains fills the rice paddies to create a kind of aquarium; this system has enabled farmers to grow two crops of rice a year for centuries. They do this using a unique form of cooperative agriculture that enables farming to flourish despite water scarcity and the constant threat of disease and pests. Rice planting and water allocation is coordinated by *subaks*; these bring together all of the farmers who share water from a single source – such as a spring, or an irrigation canal. The *subaks* adjust cropping patterns cooperatively in order to achieve fallow periods over sufficiently large areas to minimize dispersal of pests. Having studied this unique stewardship approach to a regional watershed for thirty years, Stephen Lansing has concluded that Bali's *subak* water management system is a 'coupled social-ecological system'. Irrigation in this context, he explains, is not just a matter of delivering water to a plant's roots. Because the rice terraces are hydrologically connected to each other, the farmers have had to solve a complex coordination problem: who gets to use how much water, when, and how? A complex, 'pulsed' artificial ecosystem has evolved over generations in which the allocation of water is adjudicated by a priest in a water temple. The arrangement is a dynamic one; cooperation is continuous among hundreds of farmers whose relationships span entire watersheds. 'There is a complex adaptive systems explanation for water temples,' Lansing explains, 'but also a complex cultural one. The temples are more than just a kind of mathematical device. A great deal of attention is devoted to symbolic ritual activities such as food offerings, prayers to deities, and elaborate pilgrimages.' Rituals, says Lansing, serve the regulatory function of feedback; they embody the interdependency of upstream-downstream relationships, and codify trade-offs between water sharing and pest control that are the result of a very long-term trial-and-error process. Rituals materialize ideas accumulated over centuries.[26]

In the 1970s, this system was badly disrupted. Indonesia, which was struggling to meet the costs of importing rice to feeds its growing population, perceived a need to improve agricultural productivity. The country turned for help to the Asian Development Bank (ADB), an early funder of the Green Revolution. Government subsidies for the use of fertilizers and pesticides were introduced. Farmers were made to switch to 'miracle' rice varieties, and were also pressured to disregard the traditional irrigation schedules of neighbouring paddies and plant rice as frequently as possible. The experiment almost destroyed the *subak* system within a decade. After a brief increase in productivity, crops dwindled drastically as water shortages and infestation by vermin took hold. Miracle rice begat miracle pests as a plague of plant-hoppers devastated crops. By 1974, field workers in Bali were reporting 'chaos in water scheduling' and 'explosions of rice pests'. In 1984, Lansing told the ADB that these problems were linked to disruption of the traditional system of water management, and that their high technology and bureaucratic solutions had proved to be counterproductive. ADB officials remained sceptical until Lansing, working with ecologist Jim Kremer, built a computer model. The model, which simulated different cooperation strategies among farmers over long time periods, confirmed that coordination – from farmer level up to the level of the watershed – was best made at the temple level. Eventually, Indonesia's officials became such converts to the water temple system that now, thirty years later, *subak* has been accredited by UNESCO as a 'cultural landscape' of world importance. '*Subak* brings together the realms of the spirit, the human world and nature,' states the UNESCO declaration.[27]

FROM CLOCK TIME TO ECOLOGICAL TIME

In Bali, a too rigid scientific approach nearly destroyed a thousand-year-old system that worked – but, just in time, the wisdom accumulated over generations is regaining respect. That wisdom teaches us that nature is not a machine. It is a complex of living systems, including social ones, whose cycles operate at different speeds that are determined by a multitude of different contexts. Natural time does not progress in straight lines; it moves in cycles that are shaped by the unique qualities of different locations. People who have lived in a place for generations, like Bali's rice growers, understand this in ways that the mindless logic of globalization cannot possibly do; there is no place in a world of pre-programmed growth for the complex temporality of plants, animals, and ecosystems. The Balinese, explains Stephen Lansing, think about time in terms of the multiple, concurrent, and interlocking cycles found in nature. Their master calendar plots the rice cycle; it contains 210 days, the growth cycle of Balinese rice. A market 'week', which is three days long, has not changed for a thousand years.

The lesson of the Balinese story is not that indigenous knowledge is a superior alternative to scientific knowledge; we need both – but not in a hierarchy with science on top. If the rest of us are to work with nature, and not against it, we, too, must subjugate machine time to organic, ecological, and even geological tempos. The natural systems that sustain us move at a slower rhythm than today's economy does; feedback loops in nature are slower than the lightning-fast synapses we've built into our machines.

Different ways of knowing, as I write about more in Chapter 10, are an important factor as we make this transition. For farmers in Bali, their music is a form of ecological knowing. As explained by the musicologist Judith Becker in 'Time and Tune in Java', Gamelan music helps people connect with the multiple cycles of nature.[28] Like nature, Gamelan music is composed of multiple interlocking cycles, subdivisions of cycles, and concentric cycles; these all rotate simultaneously within each other. The experience of order appears if the cycles are integrated well, Becker explains; if they are not well integrated, the result is unsettling. I have no idea what the equivalent of Gamelan music might be for the watershed where you live – so go find a composer and get started.

WATER WISDOM

I hope I have persuaded you, in the stories above, that even small actions can have big and positive outcomes for a watershed when enough people take them in concert.[29] The action can be to design a rain barrel, help clean up a creek, map the ecological resources of a watershed, plant water hyacinths, or compose the music score of an ecosystem. If we are to transform watersheds at an ecosystem scale, a variety of different actors and stakeholders – formal and informal, big and small – need to work together. This will seldom be easy, and because every city and its watershed is unique, there is no global blueprint to follow. These challenges are daunting – but bringing watersheds back to life is a vision with immense cultural power.[30]

4 DWELLING: FROM DEPAVE THE CITY, TO POLLINATOR PATHWAYS

It's early September. Any day now, the family of swallows that have spent the summer in the eaves outside my studio will head south for the winter. Most of them will follow the west coast of Africa to avoid the Sahara; a few may travel further east down the Nile Valley. They'll take it easy at first, stopping every few miles to build up their fat reserves, but then they'll speed up. In four months, as Christmas beckons here in the north, they'll reach their destination: Botswana, Namibia, or South Africa. After just two months gorging on insects, they'll begin the epic return journey. The strongest among them will make it back in just five weeks, travelling 320 km (200 miles) a day (they're keen to get back to their nesting site before some other bird nicks it). And I thought *my* air travel was profligate.

As an artefact, the swallows' nest is hardly the Taj Mahal; it's a ramshackle structure, made of mud pellets and straw, that's stuck crookedly to the wall. But it seems to suit them well – or rather, the surrounding *habitat* does. Their physical abode is a safe place to rear their young, but what brings the swallows back every year is the environment as a whole: open air for easy flight; fresh water from the river; flying insects to feed themselves and their offspring. As the swallows twitter excitedly overhead, I envy how lightly they manage to live. I compare their external energy needs, which are tiny, to the prodigious energy flows and billions of tons of resources, gathered from faraway lands, that keep our cities going; the elevators in our skyscrapers; the huge pumps that supply our water and air, and keep our subways dry; the industrial food in our shops; the water in the taps.

Preoccupied by this contrast between our lifestyles and those of the swallows, I posed the following question to a meeting of housing association managers: do we really *need* to build more boxes? Is it beyond our creativity to provide our fellow humans with shelter and sustenance without covering more of the world in concrete? To be candid, I did not expect an easy ride from this group of experts. Their daily job is to manage 2.4 million housing units across the Nordic countries, and demand for more is unremitting. But I was surprised: many of these professionals shared my concern at the baleful influence of the Real Estate Industrial Complex. Manufacturing boxes may be good for GDP and construction firms, some agreed, but we can surely meet the social need for shelter in ways that improve the habitat, not wreck it.

TOP-DOWN NATURE

My surprisingly friendly reception in Reykjavik was symptomatic of a profound change in the ways we think about cities. The first signal of change that I noticed was in 2009 when Nicolas Sarkozy, when he was French president, asked ten architects to dream up 'the world's most sustainable post-Kyoto metropolis'.[1] The architects duly dreamed. One proposed to build economic 'buds' in an 'archipelago' around the capital. Another proposed to double the number of forests, and bring vegetable fields to the city's outskirts. A third proposed to cover up roads and railway lines with huge green canopies. Two Italian architects proposed to lay the city out as a 'porous sponge' in which waterways would be given pride of place. In this contest of the metaphors, 'sponge' struck this observer as the strongest one to emerge – but, at the time of writing, none of the competing plans has gained traction. The only project to be approved, in 2011, was a

US$30 billion public-transport network called Le Grand Huit – and the word 'sponge' does not appear in its plan.

The failure of Paris to embrace its future as a sponge, in the short term at least, did not deter my adopted country's dreamers. In 2010 a vast exhibition called 'The Fertile City: Towards an Urban Nature' explored nature in the city from multiple perspectives: historical, social, cultural, botanical, and ecological.[2] A gigantic promenade led the visitor to sixteen projects from all over the world – from Beirut to Buenos Aires. 'The Fertile City' was an engaging spectacle, but most of its luscious images featured rich urbanites lolling around on green playgrounds. Life in the Fertile City seemed, literally, to be a picnic. It was like a still life fruit-bowl painting: decorative, but hardly nourishing. The show said nothing about the social and economic changes – not to mention the mud, and the work – that would be needed for our cities to become fertile sponges in real life.[3]

Another ideas project, this one for Chicago, is based on the concept of 'Eco-Boulevards'.[4] Shocked by the discovery that Chicagoans waste over 3.8 billion litres (1 billion US gallons) of the water they extract from the Great Lakes each day, Martin Felsen and Sarah Dunn proposed to transform existing roadways, sidewalks, and parks, which comprise more than a third of the city's land, into a holistic, distributed, bio-system for recycling water. Long strips of publicly owned land would be transformed from grey infrastructure – roadways and sidewalks – into an interconnected network of parks, wetlands, preserves, bio-conduits, and native landscapes. Natural bioremediation processes – the use of plants to clean up polluted soils – would remove contaminants from storm water and wastewater. Thus cleaned, the water would be returned to the Lakes, closing Chicago's water loop. If realized, Chicago's Eco-Boulevards will remake the city in the image of its own motto, '*Urbs in Horto* – City in a Garden'. They remain, at the time of writing, a concept but a resonant one.

Undaunted by the failure of these visionary projects to gain traction, the city of Bordeaux launched a live project called Bordeaux 55,000 to realize 'urban nature' in practice.[5] Five multidisciplinary teams were invited to explore how to transform 55,000 hectares

(136,000 acres) into natural areas. They were asked to 'restore natural and human capital', find ways to rehabilitate polluted soils, and use the design principles of permaculture in their proposals. To meet this brief, the teams assembled a small army of specialists: experts in architecture, geography, economics, agronomy, ecology, planning, development, landscape, sociology, tourism, hydrology, philosophy, history, philosophy, and storytelling. Although the French project is adventurous, Bordeaux 55,000 (which is ongoing) sits firmly within a resource-intensive development paradigm. The city's 'nature programme' accompanies a vast investment in high-speed rail infrastructure designed to transform Bordeaux into the 'crossroads of south-west Europe'. The plan is grand, but not yet grounded in the realities of resource constraints; it is likely to grind to a halt as energy descent kicks in. For a more representative picture of how energy-stressed cities may evolve, one has to look elsewhere.[6]

WILD CITY

In a project called Wild City, the Rotterdam-based research group Stealth set out to understand the seemingly chaotic and non-planned processes that transformed Belgrade during the turbulent 1990s.[7] They found that many functions and services previously provided by institutions were reinvented by an accumulation of individual initiatives in trade, housing, and other public services. The most radical reconfiguration of the city, Stealth found, happens when different kinds of street trade enter and reshape public space.

By 2020, according to the Organisation for Economic Co-operation and Development, two-thirds of all the workers of the world will be employed in the informal economy. In his book *Stealth of Nations*, the American writer Robert Neuwirth describes in vivid detail how this shadow economy is reshaping our cities as the formal economy stagnates.[8] The shadow system is most dramatically visible in places most of us never see: in the vast open markets underneath the Superhighway between Shenzhen and Guangzhou; in the tri-border area of Paraguay, Brazil, and Argentina; among the derelict remains of a once popular resort on the fringes of Buenos

Aires; in a former container depot outside Odessa; on no-man's-land along the former Iron Curtain. In each of these areas informal markets occupy a shifting mosaic of small locations. For Helge Mooshammer, who is also researching the phenomenon, these informal zones mark a radical shift in urban organization – 'from geographically fixed territories to a networked ecology of filters and channels'.[9]

INFORMAL FOOD

The lightweight hardware of these informal markets contrasts starkly with the heavyweight systems that supply our cities with food and water. As I will explore further in Chapter 5, the industrial system that keeps cities fed consumes ten times more energy running itself than it delivers as nutrition that you and I can eat. These food systems are only viable when fossil fuels are abundant. In a visceral response to this fragility, informal food and water systems are sprouting up everywhere, too. Five years ago, urban farming was a fad. Today, serious urban agriculture projects have taken root in dozens of northern cities. A powerful grassroots movement has given us community-supported agriculture and box-schemes, the 100-mile diet, and Slow Food. Sales of vegetable seeds have skyrocketed; backyard chickens are now commonplace; and schoolyard gardens, organic farms, and farmers' markets have proliferated. Together with thousands of grassroots projects to capture rainwater and depave hard surfaces, the informalization of food and water flows is transforming urban-rural relationships, and with them the metabolism of cities.

A few cities have already adapted their approach to planning to accommodate these developments. In the US, the city of Portland commissioned a report called *The Diggable City* to determine which of its properties might be suitable as community gardens or other kinds of agricultural uses.[10] Having identified 289 locations, the report yielded an action plan for the modification of land tenure, access to water, level grade, site security, and other considerations. Chicago set up a project called NeighborSpace to serve as a land management intermediary between the city and community groups seeking to develop projects on vacant public land.[11] Social practices are a key part of this urban transformation. Urban farming is as much about the design of ways to share and collaborate, as it is about what to put in the beds. New services, policies, and infrastructures are needed to support food co-ops, collective kitchens and dining rooms, community gardens, cooperative distribution platforms, seed banks, hothouses, nurseries, and other enhancements of community food systems.

SUBURBIA, TOO

The American writer James Kunstler described suburbia as 'the greatest misallocation of resources in the history of the world'.[12] Which may be true, but what's done is done – and besides, most of us have no alternative but to adapt to the places we live in now. On the positive side, although sprawl is frequently blamed for environmentally damaging transport intensity, the collapse of communities, and even obesity, suburbanites do also live on pockets of land in communities that have not been paved over with asphalt and concrete. Much of this land is subdivided into manageable parcels that could be productive once lawns are removed and the soils restored. For Bill Mollison, one of the founders of the permaculture movement, there's enormous potential to transform suburbia into a semi-agrarian patchwork of communities for localized food self-sufficiency. His vision is helped by the fact that some of the most widespread landownership in any mode of settlement is to be found in the world's suburbs.[13] A lot of land with potential for food growing will have to be depaved – but as car use declines, the potential areas available will be huge. Roads and parking facilities typically cover more than 50 per cent of urban land in most city centres and shopping malls; there are up to two billion parking spaces in US cities alone. These impervious surfaces prevent rainwater from entering the soil, instead diverting it to nearby waterways. This water carries pollutants such as oil, antifreeze, plastics, pesticides, and heavy metals from the roads into local streams and rivers.[14]

The liberation of soils from their imprisonment by parking spaces began as an illicit form of activist action. Small groups of guerrilla de-pavers, wielding pickaxes and wheelbarrows, would remove hard surfaces to reveal the underlying soil bed. Their actions were followed in 2005 by a more mainstream art project called PARK(ing) Day in which citizens transform metered parking spaces into micro-gardens. What began as a two-hour installation in a single parking bay has now become a global movement; in 2012 the action involved more than a thousand parking bays on six continents.[15] Depaving went mainstream with the establishment in Portland of Depave, a non-profit organization that promotes the removal of unnecessary pavement from urban areas to create community green spaces and mitigate storm-water run-off. Many of suburbia's paved spaces and roadways could also be repurposed to gather the water we would need to grow food for ourselves.

Cost is a powerful incentive for depaving. It costs US$215 per square m (10 square ft) to build a road, but just $54 to open up the space for an urban farm. In some country districts, budgets cuts are doing more for the depave movement than activism. Because the cost of rebuilding an asphalt road has more than doubled over the past ten years (largely because asphalt cement is a petroleum-based material) some US counties are opting to pay about $2,600 per 1.6 km (1 mile) annually to maintain depaved roads as against about $75,000 per 1.6 km (1 mile) to reconstruct them using asphalt.[16]

MAKE ROOM! MAKE ROOM!

Can depaving seriously free up enough land to feed a booming population? In the European Union alone, about 1,000 square km (385 square miles) of land, an area larger than Berlin, are being taken each year for housing, industry, roads, or recreational purposes.[17] Reversing that trend and restoring concrete-covered land to fertility is an enormous task – but not an impossible one. Ugo Bardi tells the story of a couple in Turin, Italy, who decided to give a patch of fertile land to their children as a gift.[18] They obtained it by demolishing a row of concrete garages they had inherited. It was a lot of work; the concrete had to be cut and broken to pieces, and the rubble carried away. Restoring the fertility of the soil took truckloads of dirt, charcoal, and other nutrients. The process was slow, messy, and expensive.

But it was also a subversive idea – an idea with the potential to spread. Until now, developing land has meant building on it; that's how you make money. Destroying property to restore fertile soil is something that nobody in her right mind would normally do, marvels Bardi, but someone did it. The end result was a patch of fertile soil where grass and flowers grow. Just a few tens of square metres, not much in comparison to the trillion remaining to be recovered. But it is a first step!

OVERBUILT

Think back to the swallows I mentioned earlier. Many animals and birds meet their needs like early man used to do: they hunt, and they gather. Without pretending to mimic swallows or hunter-gatherers literally, we can surely take inspiration from their spatial intelligence and ask: how many buildings do we need anyway? For Fits van Dongen, state architect of The Netherlands, the answer is: none. Van Dongen stunned his colleagues and the construction world in 2013 when he called for an end to all new building. 'We have half a million square metres of office and industrial space, and 30,000 homes standing empty,' explained van Dongen, before proposing that soft green infrastructures, such as spaces for urban agriculture, were a better priority.[19]

The realization that rapid urbanization is a symptom of systemic problems, not a solution, is taking hold in newly industrializing countries, too. In 2014, when India's new government announced plans to build the 100 'smart cities' I mentioned in horror in Chapter 1, the lack of enthusiasm for the proposal was startling; critics pointed out that 25 per cent of India's total primary energy demand was already being used to manufacture building materials and this building binge would exacerbate the country's already acute energy crisis.[20] Even potential clients for new buildings are asking hard questions about the need – or otherwise – for these capital-intensive assets. Most of the fastest-growing 'church-planting' movements, for example, are multiplying without the need for custom-built churches as physical buildings. For Rick Warren, the entrepreneurial pastor, 'massive building programs are often a waste of money; they're a barrier to exponential growth'. Warren reminds his followers that the period of fastest growth for Christianity was during its first 300 years – when there were no church buildings at all.[21] A similar lesson is permeating the business world. For Bill Mayon-White, a professor at the London School of Economics, the physical assets owned by most corporate giants represent 'an albatross hanging around their necks'. Companies gain flexibility by not owning physical assets, by concentrating on ownership of intellectual property and moving that around.

The most valuable work in today's urban economies benefits from close proximity – and a city's streets, interspersed by shared workspaces, can support that kind of economic activity more cost effectively than buildings; simple technology platforms can render almost any urban location a potential worksite. In Mumbai, the urban research group Crit has developed a diagnostic tool to measure what it calls 'transactional capacity' of a city. Based on its findings, Crit deploys social innovation tools to enhance the complex landscape of services, relationships, and physical spaces that make a city truly smart.[22]

Even as we stop building unneeded new buildings, many empty ones will still remain – and what to do with them? In *The Economy of Cities* (1969), Jane Jacobs predicted that cities would become 'the mines of the future' – and not just metaphorically; the 'standing crop' of industrial metals already embodied in artefacts among the ten richest nations would require more than sixty years' production of these metals at 1970 rates.[23] Four decades after she made it, Jacobs's forecast is coming to life amid a proliferation of grassroots projects to dismantle buildings, reclaim tower blocks, and renew trailer parks. Disused buildings often contain valuable amounts of embodied energy, so the deconstruction or 'decon' trades are flourishing. A Deconstruction Curriculum in the US trains people to be 'green workforce ready'. Eight-week decon courses may be taken in Fort Dodge, Sioux City, Spirit Lake, and Cedar Rapids.[24] In Savannah, Georgia, building materials reclaimed from a public housing demolition project are being diverted to a Girl Scouts eco-camp, public school playgrounds, and community gardens.[25] In the UK, more than 60,000 people have joined a decon resource hub called Bricks and Bread that opened in 2009.

Its founder, Trudy Thompson, went on to create a social franchise model to replicate its systems and know-how; more than 300 franchisees are now using her innovative methods.[26] For its part, the UK's Asset Transfer Unit helps communities take over underused land or buildings – and do so legitimately.[27]

PATCHWORK CITY

The lesson of the decon boom is that rewilding cities is not much about the creation of wide open spaces; it's more about patchworks, mosaics, and archipelagos. When parks were built in past centuries they were called the 'green lungs' of towns. Decades of oil-fuelled overdevelopment has put an end to those expansive days – but a new generation of 'greening designers' have abundant man-made assets to work with. There are parks, cemeteries, watercourses, avenues, gardens, and yards to adapt. There are roadside verges, green roofs, and facades to plant. Sports fields, vacant lots, abandoned sites, and landfills can be repurposed. There are large and growing numbers of abandoned buildings and ruins, empty malls, and disused airports to modify – not to mention the abandoned aircraft that, before too long, will be parked there.

In Vienna, a design firm called Biotope City develops 'micro green spaces' to transform neighbourhoods. In the densely developed Haslingergasse district, for example, the group covered the walls, balconies, and ledges of 150 social housing blocks with greenery.[28] Thanks to the participation of local schoolchildren, nesting boxes for birds and insects were also added. A similar patchwork approach is emerging in the Jæren region of Norway whose landscape has been battered by the footprint of the oil economy. Undeterred, the architect Knut Erik Dahl teaches young designers to look for and appreciate the tiniest examples of biological life: solitary plants, rare lichen, and insects in among the people, goods, and buildings. Students make large-scale maps of each location on paper, by hand. It's low-cost, hands-on work. They call it 'dirty sustainability'.[29]

This new approach is all about nurturing patches, some of them tiny, and linking them together.[30] Think about 'my' swallows.

They thrive in open habitat, not in a closed-off property. Poor people, who live in tiny private homes compared to rich people, know this already. The average floor space per person in low-income countries is about 8 square m (80 square ft) compared to 60 square m (600 square ft) per person in the US – a nearly eight-fold difference.[31] But Mcmansions can be lonely places and the trick to thriving in small dwellings is access to shared spaces and services. The secret lies in the connections. When habitat patches in the landscape become fragmented, biodiversity suffers. The same goes for humans. Habitat fragmentation is an issue in cities, too. Ecologically minded landscape designers therefore try to ensure that valuable patches do not become too small or isolated to support species. In France, several cities are working to establish 'ecoquartiers', or eco-neighbourhoods, in otherwise gritty urban contexts. The idea is to 'seed' functions such as rainwater capture, or Sustainable Urban Drainage Systems (SUDS), in the hope that they will propagate and spread without ongoing support from city authorities – or budgets.

LEARNING FROM THE SOUTH

Many cities in sub-Saharan Africa are already 'green' in a bioregional context.[32] Cityscapes in which houses are surrounded by food growing are normal. The 'peasant economist' Teodor Shanin, who reckons that the informal economy supports three-quarters of the world's population, reminds us that 80 per cent of all farms in the world – 445 million of them – occupy 2 hectares (5 acres) or less – many in or near cities. From Lima in Peru, to Kinshasha in Zaire, a lot of expertise has accumulated on how to manage the complex social-ecological networks of urban and suburban agriculture. The main lesson: necessity, more than aesthetic reverie, is the mother of urban transformation.

Does city-grown food make a meaningful difference? A recent US study suggests that it does – or at least can. Researchers explored what it would take for Cleveland – a Rust Belt city with lots of potential green space – to feed itself. The results were startling: in one scenario, the use of 80 per cent of every vacant lot generated

22–48 per cent of the city's fruits and vegetables, 25 per cent of its poultry and eggs, and 100 per cent of its honey. If commercial and industrial roofs were added to the equation, the city could provide up to 100 per cent of its needed fresh produce, 94 per cent of its poultry and eggs – and 100 per cent of its honey.[33]

Growing food is one thing; storing it is another. Awareness of energy descent is reviving interest in the ways we preserved food in the past. For thousands of years, people in different cultures have preserved foods and vegetables using lacto-fermentation; in this fossil-fuel-free process, lactic acid acts as a natural preservative to inhibit putrefying bacteria. What began as a foodie fad is fast becoming mainstream as more people seek out practical ways to conserve food without resort to refrigeration or chemicals. Right next to City Hall in Seoul, in Korea, I visited a farmers' market in which fermented fish and vegetables were displayed in rows of tubs; these were interspersed with large mounds of bright red kimchi paste. Unlike canning and pasteurization, which kill all microorganisms in a food, fermentation takes the opposite approach: it promotes natural life in the form of bacteria and yeast in which microorganisms harmful to humans cannot live. In Korea, kimchi pickles are stored in *onggi*, or earthenware jars, which are shared and looked after by the community as a whole.[34] As similar tubs and jars crop up in northern cities, too – along with neighbourhood bakeries and breweries – a growing number of citizens have become unwitting experts in biological symbiosis.[35]

POLLINATOR PATHWAYS

Urban patchworks as a living infrastructure for bread and beer production are also friendly for pollinating insects. With bee and insect populations plummeting across the US and Europe, small gardens have enormous potential to act as archipelago-like nature reserves. England's 15 million urban backyard gardens, for example, occupy more ground than all of the country's official nature reserves combined.[36] To link these micro-sites together, and thereby to strengthen a city's natural food web, so-called pollinator pathways are being developed. In Seattle, for example, the artist and ecological designer Sarah Bergmann established a mile-long series of planting strips along Seattle's Columbia Street to create a corridor between two green spaces at opposite ends of the city. Each planting strip – usually a band of grass between sidewalk and street – is transformed into a pollinator-friendly garden that offers viable food and habitat to vitally important insects.[37] In the UK, a nationwide project has started to establish 'Bee Roads' across the country; these will act as food-rich main routes for pollinators.[38]

Filling up cities with plants and trees has energy-related benefits, too. Cities, and not just rainforests, can provide ecosystem services. When researchers visited parks, golf courses, abandoned warehouses, and household gardens around the city of Leicester, they discovered that urban vegetation stores ten times more carbon dioxide than previously assumed.[39] Mature trees provide a canopy of shade to help cool buildings and streets; this reduces energy usage. Street plantings can also entail the removal of concrete, and mulched trees help rainwater soak into the soil underneath depaved surfaces. Trees also help prevent soil erosion; save water; shield children from ultraviolet rays; provide food; help sick people get well; reduce violence; and provide a canopy and habitat for wildlife. A recent Portland study suggests another surprising benefit: healthier newborns. Researchers found that pregnant women living in houses graced by more trees were significantly less likely to deliver undersized babies.[40] (Whether larger babies go on to consume more resources was not reported.)

A pioneer in urban forestry projects, and of novel forms of social organization to create them, is Andy Lipkis, founder of TreePeople in Los Angeles in 1973. TreePeople's Citizen Forester programme organizes volunteer tree plantings and tree care events along city streets and in neighbourhoods throughout Los Angeles County. TreePeople has also distributed thousands of fruit trees to low-income communities; these become functioning community forests – in backyards, on school campuses, and in community gardens. Planting trees is just one aspect of the group's work. For at least five years after a tree is planted, its progress is monitored by

Tree Care Coordinators who organize tree care events as needed; this dramatically increases the chance of the tree's survival to maturity.[41] Lipkis's pioneering work depaved the way for the creation, in 2005, of a Sustainable Urban Forests Coalition. This platform brings together city planners, educators, landscape architects, non-profit leaders, scientists, arborists, foresters, nurserymen and women, and many other professionals who care for, monitor, and advocate for trees and our urban forests as a whole.[42] The American Forestry Service then established an Urban and Community Forestry website whose pages are filled with happy-making stories. A manual filled with best-practice case studies has been distributed to more than a thousand planning agencies.[43] In Europe, a Forum on Urban Forestry collates discussions on how best to manage old urban forests, parks, and cemeteries. The island state of Singapore plans to transform itself from a 'garden city' to a 'city in a garden' by reforesting the entire city state. Tai Lee Siang, president of the Singapore Green Building Council, laments the fact that town planners, when they start drawing, almost inevitably draw the roads first. Siang plans to reverse that sequence. Quoting a NASA estimate that there are about 57 trees to each person on Earth, he announces a new target for Singapore: a ratio of 100:1.

REWILDING THE CITY

As people get reacquainted with real physical work in our cities, we are learning a special respect for solutions evolved by nature over the last 3.8 billion years. As Janine Benyus reminds us, other life forms than our own are able, expertly, to move water, capture the Sun's energy, provide shelter, store food, recycle nutrients, share resources, build communities, control population, and manage ecosystems – all without human intervention. We've exhausted ourselves trying to control our environment using fossil-fuel force – only to be reminded that ecosystems can manage this effortlessly on their own.

This prompts an interesting question: pull that weed out of its crack in the pavement – or let it grow? A growing number of people are inclined to welcome back the weeds and let nature take its course.

Seen through the lens of biodiversity as wealth, the city turns out to be richer than we thought. There is more biodiversity in many cities, for example, than outside them. In the 'country', industrial agriculture blankets large areas with monocultural single crops; among 20,000 species of edible plants in the world that we know about, fewer than 20 species now provide 90 per cent of our food. Some cityscapes, in contrast, are far more diverse. Urban biologist Claudia Biemans, a local naturalist and edible plants researcher in The Hague, has identified about 300 different species in 1 square km (⅓ square mile) of her city; this compares to 50 different species found in the same area of managed countryside nearby. 'Bees know this very well, and are more to be found in cities these days,' she points out. On walks called 'Stalking the Wild', Biemans guides people to ecological niches in the city where plants don't just survive, but thrive.[44]

Much of this urban nature is edible. Herbal fruits, leaves, and edible flowers grow on walls and roadsides, between paving stones, and in other untended spaces. Lynn Shore in Amsterdam, trading as Urban Herbology, is among a growing band of urban foragers who help citizens find herbs, use them in cooking, and learn about medicinal preparations. Shore's activities include seed and plant swaps, urban herb walks, and 'gatherings for urban herbies'. In Los Angeles, a so-called 'rock star of foraging' called Pascal Baudar has turned his passion into a thriving business; Angelinos pay $100 a session each to join his 'Gourmet Foraging Sunset Experiences' on which they learn about the culinary uses of weeds found in the local landscape. Baudar's wild food classes sell out weeks ahead.[45] Less prosperous foragers, the majority, are using a free mobile phone app called Boskoi to map the edible landscape; in an activity called 'augmented foraging' they share the location of wild food in public space. The word 'Boskoi', say its Dutch developers, is taken from Greek; it dates back to the tradition of desert hermits in the South Egyptian and Sudanese desert. This hardy band survived exclusively on wild herbs and rainwater, and were said to graze with wild herds of cattle. A splendid model for the post-peak city, then.[46]

Alerted by these innovative projects to the fact that cities support about 20 per cent of the world's bird species and 5 per cent of its plants,[47] a growing number of city fathers now realize that their cities are dynamic nodes of biological activity in their own right, and that ecosystems in 'their' bioregions are a source of value.[48] In the UK, brownfield sites contain more rare insects than do ancient woodlands and chalk downlands.[49] And the US Forest Service, once sceptical that anything urban could be wild, now supports a growing urban forest programme.[50] In China, too, 600 million volunteers planted 64 billion trees in 2012.[51] Urban ecology and urban wildlife programmes are proliferating on university campuses.[52] Boasting that 'my city has more wildlife than yours' suddenly sounds so much better than banging on endlessly about its creative class.

Claudia Biemans, who I mentioned above, is among a growing community of researchers for whom the evolution of plants and animals in cities is an urban design opportunity. In New York, scientists have identified mutations in more than a thousand genes in the city's mice – far more than in mice from out of town. Extinctions, invasions, and adaptations have created a complex mixture of native and non-native life forms. Not all of this change has been positive; Manhattan was once home to twenty-one native species of orchids assumed to be extinct due to the replacement of woodland by open urban spaces.[53]

Or are they really extinct? Their seeds may still be there. The notion that older ecologies lie beneath our cities, waiting to self-resurrect, has long fascinated artists – and now scientists, too. Paleobotanists have discovered that 1 square m (10 square ft) of urban soil can contain tens of thousands of seeds that persist in a state of suspended animation, waiting to be woken from their slumber. In his essay 'City of Seeds', the writer Daniel Mason reflects that a two-and-a-half centuries old tradition of urban botany has yielded a startling insight: the flora of the city 'is essentially a flora of the city's destruction'.[54] Unlike the managed green of parks and gardens, which only grow in pockets of protected isolation, the wild plants of a city need 'the cracks, the pavement split, the palace abandoned'. Beyond the managed gardens and the wild invaders of our roads. Mason concludes, is 'a hidden, potential flora, an *idea* of a forest, not in competition with the city but existing alongside it, patiently, waiting to become manifest'.

Although their activities are separate, these artists, writers and scientists agree that the city is part of nature, not outside it. 'Instead of manicuring sustainable gardens,' states Urban Cannibalism, 'we celebrate the spontaneous surplus of the city's life.' This group does not restrict itself to plants; it proclaims itself to be a conduit for 'the organic and inorganic voices of the city, of the liquid flows of minerals and invisible ecologies of microorganisms that constitutes the bodies of buildings and beings.' Buildings breathe and ferment, too, they explain – just very slowly; even the hardest wall is host to the invisible food chains of microbes and mould. The closer you look, the more blurred becomes the border between organic and inorganic life.[55]

PLACE MAKING

I'm finishing this chapter outside the Portakabin control room of Shambala, a summer festival in England. On the wall is the street plan of what looks like a mid-sized town. Fifteen thousand people have indeed filled a vast field with tents, yurts, sound stages, composting toilets, drinking water tanks, hot tubs, food vans, charging stations, yoga enclosures, a barber shop, a meadow filled with aromatherapists, vending machines in a caravan, and pagan circles around wood-burning stoves. Surrounding Shambala's central core is a densely packed suburbia of tents; in these the sleeping area per person – a couple of square metres – is similar to the space available to billions of people in the world's other favelas.

Most of Shambala's prosperous urban tribe will return to a world of concrete and media when the festivities end – but, for two-thirds of the world's population, nomadism and contingency are now everyday conditions of life. Most of the world's 800 million urban farmers, for example, grow food because they need to eat it, not to be cool. In megacities across the Global South, informal settlements are also filled with the pop-up retail, food trucks, street traders, guerrilla gardening, and informal parks, that – at Shambala – are celebrated as fashionable novelties. In the world's refugee camps and post-disaster settlements, too, a dynamic variety of social micro-economies enables people to share energy, materials, time, skill, software, space, or food. These activities depend more on social energy, and trust, than on fixed assets and real estate. There's an emphasis on collaboration and sharing, on person-to-person interactions, on the adaptation and reuse of materials and buildings.

These resource-light ways to meet daily life needs are usually described as poverty, or a lack of development. But in thirty-five years as a guest in what used to be called the 'developing' world, I've realized that people who are poor in material terms are highly accomplished at the creation of value in ways that do not destroy natural and human assets. DIY-urbanism, in other words, is second nature for people who cannot depend on the high-entropy support systems of the industrial world. This is not to trivialize the extreme challenges faced by poor people on a daily basis; but, to the extent that a regenerative economy is based on local production, human labour, and natural energy, the poor people of the world are further along the learning curve than the rest of us.

What the Earth needs, and what the Real Estate Industrial Complex needs, are two different things. The world is overbuilt. As our measure of economic progress shifts to health of the soils, and biodiversity, the practical focus of our efforts is shifting to a city's inhabitants – including non-human ones – and to ways of improving habitat for them all. The writer Thomas Berry described as the *ecozoic* this 'reintegration of human endeavours into a larger ecological consciousness'. It's the right way to think about, and act, in our cities.

5 FEEDING: FROM SOCIAL FARMING, TO FOOD AS A COMMONS

Not long after my soil-making day in the Cevennes mountains, I found myself in the centre of the city of Carlisle, in the north-west of England, at 7 a.m. The railway station forecourt was empty except for a large truck whose driver was unloading packaged food into a cafe. An incredible, raw-edged roar of noise came from the refrigeration unit on top of his cab; it was so loud that I couldn't hear a word when someone called my mobile phone, so I retreated into the station cafeteria. It was little better inside: two refrigerated drinks machines were roaring so loudly that the sales assistant had to shout to tell me the price of a coffee. Getting off the train in London a few hours later and shopping for a snack in a branch of Marks & Spencer, I was aware of another, even louder, background roar. Curious about its source, I counted out 78 m (256 ft) of chiller cabinets in that one, not-so-big, urban food store. As I stepped out into the busy street, the noise of the chiller cabinets merged with the roar of the traffic – and in more ways than one. A lot of that traffic noise turns out to be food-related, too: agriculture and food now account for nearly 30 per cent of goods transported on Europe's roads; in the UK, 25 per cent of car journeys are to get food. We increasingly eat food while moving, too: 70 per cent of fast-food sales in the US are at the drive-through window.

Faster, fatter, fatal. If food sounds bad, it probably is bad – but it's the silent and unseen costs of industrial food that do the most damage. Poor diet and physical inactivity account for 35 per cent and rising of avoidable causes of deaths in the US,[1] but the food-driven obesity pandemic is not confined to the North. In New Delhi, a third of school-age children are obese – mainly because the sugar content of their diet has risen 40 per cent during the last half-century, and its fat content by 20 per cent. That's 'development' for you. Bizarrely, municipal authorities in Delhi want to ban the city's 300,000 street food vendors – who use very little sugar or saturated fat – in the name of hygiene and modernization. Oh, and processed food does not just clog our own arteries, it clogs our cities, too: fat deposits poured into drains by fast-food outlets are blocking sewers in cities across the United States.[2]

But back to that food-related noise. What does it signify? In my search for an explanation, I learned about a phenomenon called the 'geography of cold'. The writer Nicola Twilley discovered that the diet of most North Americans and Europeans depends on the existence of a vast 'distributed winter' – a seamless network of artificially chilled processing plants, distribution centres, shipping containers, and retail display cabinets that creates the permanent global summertime of our supermarket aisles.[3] This wintry journey includes foods that, on the surface, seem unlikely candidates for refrigeration: peanuts are stored in giant refrigerated warehouses in the state of Georgia, and half of all potatoes eaten in the US are frozen as French fries. The refrigerated cold chain, Twilley explains, is a key enabler of so-called 'production agriculture', with the grim consequences for poor farmers and ecosystem that we hear about every day. Noise also means wasted energy, which is why an American farm in the early 1800s would have been a much quieter place than that pavement in London. Back then, the balance between calories expended in food production and calories that we ended up eating, was about even: 1:1. Under today's system, that ratio is more like 12:1.

FEEDING THE WORLD

Because one billion people live in hunger, and world population is, indeed, growing, it's been easy for huge companies to frame the food debate in terms of productivity, and how to increase it. A proclamation by Monsanto is typical: 'Farmers must produce more food in the next fifty years than they have in the past 10,000 years combined.'[4] Proponents of genetically engineered crops insist that they will increase yields to end hunger, reduce costs, and improve the livelihoods of farmers and poor people; what they omit to mention is that these crops will be grown from seeds owned and controlled by private companies. The social costs of increased production are also glossed over. The Alliance for a Green Revolution in Africa (AGRA), for example, speaks about 'land mobility' – but does not explain that this means moving farmers off their farms to make way for large-scale mechanized agriculture. AGRA does not explain where these people will go and live, or how they will be re-employed – but most of these millions of small farmers end up in the slums of fast-growing megacities.[5] The biggest producers promote the benefits of 'climate smart agriculture' and 'sustainable intensification' with particular fervour – but they are silent about the consequences: the large-scale privatization of land, the decimation of peasant economies, contract farming, and increased chemical inputs. The language used by management consultants is particularly soporific and reassuring. Words like 'liberalization' 'opening up' and 'opportunity' were used by McKinsey to promote the idea that India, a country of 500 million small farmers, should become the 'food factory of the world'. The devastating costs to social and environmental quality, and public health, were not mentioned by the men in dust-free suits.[6]

Politicians, with their inherited fear of hungry crowds with pitchforks, have long been persuaded that only BigAg can feed the world – but a stream of scientific studies is dissolving that consensus. Hunger is a distribution problem – not scarcity of food, nor surplus of people. For lawmakers, their best defence against pitchforks lies in support for the world's family farmers who, after all, produce 80 per cent of the world's food on just 24 per cent of the world's farmland.[7]

According to one respected non-profit, Grain, if the yields achieved by Kenya's small farmers were matched by the country's large-scale operations, the country's agricultural output would double; in Central America, the region's food production would triple; and if Russia's big farms were as productive as its small ones, output would increase by a factor of six.[8] For Olivier de Schutter, Special Reporter on food for the United Nations, the eradication of hunger and malnutrition is an achievable goal if we help small-scale food grow in ways that leave the land healthy, and sell most of the produce locally without having to be dependent on large buyers.[9]

Turning the tide will be tough; the share of productive land held by small-scale farmers is shrinking fast. From Kenya and Brazil to Ethiopia and Spain, rural people are being displaced, threatened, beaten, and even killed by a variety of powerful actors who want their land for large-scale production. In the last fifty years, Grain reports, a staggering 140 million hectares (346 million acres) – the size of almost all India's farmland – have been taken over by four industrial crops: soya bean, oil palm, rapeseed, and sugar cane. Experts predict that the global area planted with oil palm will double in the next decade, and the soya bean area will grow by a third, if the productivist system remains unchecked. These crops don't even feed people; they feed our cars with agrofuels. Finance, not a lack of production, is another major cause of food insecurity: in a Western food shop, for every $10 that you or I spend at the checkout, only 60 cents end up with the farmer. The remaining $9.40 – the 'added value' – represents turnover and profit for the industries involved.

After that noisy truck in Carlisle first started me writing about food, the horror stories quickly piled up. I discovered that China exports billions of pounds of tomato pulp – to Italy;[10] that Europe exports planeloads of frozen chicken to chicken-filled Africa;[11] that a single 'cut and kill' meat factory may now process ten thousand factory-reared pigs in just one day;[12] and that Coca Cola and Pepsi spend more on advertising in a year than the entire budget of the World Health Organization. But bad news stories on their own are dispiriting. Noise is a symptom, not the cause, of our food worries.

So, too, is obesity. A shed filled with five-weeks-of-living-hell chickens is a symptom, too. To be told that something called a 'food system' is killing us is helpful, but too abstract. I needed to find a story that showed a way to change a system in the real world.

BEYOND THE ONE-GALLON ALMOND

'This could be it.' The speaker, Dan O'Connell, is peering through a grill into the cavernous interior of a boarded-up corner shop in downtown Fresno, California. His fellow explorer, Kiel Schmidt, concurs: 'It'll take a bit of work, but we've got a bunch of people with skills lined up to help.' For Schmidt and O'Connell, two founders of an organization called The Food Commons, the building is on their shortlist for a hub and retail store that will make fresh food available to some of Fresno's 500,000 poorest citizens.[13] Within a few years they plan to open a retail hub in each of the city's food deserts – and this will be the first.

Our location certainly fits the bill of a food desert. We've driven for half an hour past miles of empty buildings and cleared city blocks on what used to be prime agricultural land; before it was built over, small farmers would grow up to 350 crops here all year round. I'm astonished to be told that this is 'downtown'. Fresno was a ground zero of the 2008 property crash, and many abandoned buildings have been bulldozed, leaving only dust. On the corner opposite, a sign on the shabby gas station promises 'Gas. Food. Liquor'. 'People around here buy most of their food from places like this,' Schmidt explains. 'There might be the odd banana, but most of it's cheap pre-packaged stuff.' Fresno's poorest citizens, I discover, although they live in the middle of the richest agricultural region in the entire world, get most of their food from these overpriced gas stations and mini-marts. One Fresno borough – zipcode 93706 – also suffers from the highest environmental vulnerabilities in the whole country thanks to the 16.8 million kg (37 million lbs) of fertilizer dumped on the soils of the Central Valley every year; local residents live with higher health risks than anyone in California.[14] Of the honeybees who visit almond plantations here, 60 per cent die.

After being hit with this Bladerunner-like guided tour, my mood lifts when Dan and Kiel take me next to one of the five sites of Tower Urban Family Farms (TUFF) in Fresno's Tower District. The compound, where Kiel also lives, is filled with trays of seedlings. In 2013, its first year, TUFF grew one ton of tomatoes, and that yield is growing by 50 per cent a year. It's a modest crop, but TUFF also supplies seedlings to the other urban farms and community gardens that are springing up across the city. We move on to a large open space that was once a US Department of Agriculture test site, but is no longer needed by the federal government; most of its 19 hectares (47 acres) now lie dormant, but families of Hmong people from Vietnam cultivate 100 small plots in a central area.

Over lunch at Peeves Pub and Market – as I drink locally brewed beer, and eat super-fresh cheese and salad – the Food Commons team explain to this shell-shocked visitor why so much social and ecological devastation has befallen some of the richest growing land in the world – and what they plan to do about it. Although the Central Valley is home to the world's largest swathe of ultra-fertile Class 1 soil, it's also a semi-arid landscape with a natural rainfall of just 500 mm (20 in.) per year. The writer Joan Didion described as the 'plumbing narrative' the illusion that California's agricultural success story would be assured by a perpetual flow of imported water[15] – but the illusion has been maintained for decades by a massive network of reservoirs, canals, and pumping stations that transfers water from Northern California to the rest of the state.[16] From the start, the project promised more water than it could ever deliver long term, and by 2014 the system had all but dried up. Industrial-scale farms that could afford the hefty costs began digging deeper and deeper to tap into once-unreachable aquifers in what local reporter Vic Bedoian described as a 'hydrological arms race'. The Central Valley is consuming twice as much groundwater as is returned by nature in the form of rain and snow, and the food industry is mining water that is thousands of years old for crops that have a life expectancy of a few years. To make matters worse, California farmers have replaced lettuce, tomatoes, and other annual crops with huge plantations of almonds, pistachios, and other high-value – but thirsty – crops. In money terms, the switchover has been highly profitable: nut production in California brings in $7 billion in sales every year. But as the money rolls in, the actual food goes elsewhere. Although large-scale producers like to say they are 'feeding America', 85 per cent of their almond production is exported. It takes 4 litres (1 US gallon) of water to produce one single almond, so the net effect is like exporting 91 billion litres (24 billion gallons – that's nine zeros) of water halfway across the world – from a near-desert.[17]

CONNECTING TOP-DOWN AND BOTTOM-UP

Undaunted by this nutty situation, The Food Commons team is taking small steps on an ambitious journey of change. Community-owned grocery stores are in the pipeline. Community Supported Agriculture schemes, such as The Farmer's Daughter, are making headway. There are small but viable farmers' markets. Community gardens and urban farming are propagating across the city. There are projects to connect schools with local farmers. Top-down too, at a policy level, chinks of light are visible at the end of the BigAg tunnel: in 2014, for example, the US Department of Agriculture announced a $78 million investment in local and regional food systems including food hubs, farmers' markets, aggregation and processing facilities, distribution services, and other local food business enterprises.[18] Cooperatives, non-profit organizations, Native American groups, and individuals are eligible. Of course, $78 million is a tiny amount compared to the tens of billions that continue to be lavished on production agriculture – but it's a start.

Knowing the list of ingredients – and laying them all out on a table – is not the same thing as making a cake. The myriad social innovations in Fresno, as in hundreds of other cities, are cheering, and exciting – but something extra is needed if we are to transform agriculture systems as a whole. This is where The Food Commons comes in. Their approach marks a radical shift from a narrow focus on the production of food on its own, towards a whole-system approach in which the interests of farm communities and local people, the land, watersheds, and biodiversity are all considered together. The Food Commons is conceived as a kind of connective tissue that links together food-producing land, ideally held in common by community trusts; support infrastructure – such as distribution and retail centres; and support services, whether legal, financial, communications, or organizational. Its co-founder, Larry Yee, describes The Food Commons as 'a whole new cloth'.

Each Food Commons consists of three components. The first is a non-profit Food Commons Trust that will acquire and steward food-growing assets such as land and physical infrastructure. These commonly held assets will be leased to participating small farms and businesses at affordable rates; and because these assets are held in perpetual trust, they will benefit everyone. A second component, a Food Commons Fund, is a community-owned financial entity that will provide affordable capital and financial services to all parties in the regional food system; this will ease the cash flow problems that cripple so many small farmers. Each region will also contain, thirdly, a Food Commons Community Corporation, a locally owned and cooperatively run hub that will connect myriad small enterprises: farms, food processors, distributors, and retailers. Support services provided at each hub will include administration, marketing, scientific knowledge about sustainable agriculture, technical assistance, and specialized vocational training.

Fresno, where I visited, is the first working implementation of The Food Commons; it already has the active support of the city's business, academic, and social justice communities.[19] Two other prototype Food Commons are in development in Atlanta and New Zealand. Once these initial proof-of-concept prototypes are up and running, Larry Yee is confident the model can be adapted in many places around the world. 'The biggest constraint right now is investment money,' Larry told me, 'but we know the money is out there.' For now, his priority is to recruit local and regional banks that are already active in the agricultural credit and mortgage business; he's telling them one regional Food Commons can be started for US$100–150 million.

BACK TO THE FUTURE

Although, at first reading, the Food Commons plan for Fresno may strike the reader as being far-fetched, I hope to persuade you that it's but one signal of a worldwide transformation in the way urban populations are fed, and that an ancient assumption of city design – that cities are for people to live and work in, and the countryside is for growing food – is being swept away by necessity. For one thing, dozens of city-wide experiments in other countries share some or all of its features: from Rosario in Argentina to the South Bronx, Portland, Curitiba in Brazil, Freiburg in Germany, Mexico City, Zurich, and Barcelona, citizens are rediscovering how to grow fruits, vegetables, and herbs – as well as raise birds and livestock – within city limits. The sheer number of people involved is impressive: as a global average, 25 to 30 per cent of all city dwellers grow food in cities today.[20] In Havana, Cuba, 12 per cent of urban land is dedicated to agriculture; 11,000 hectares (27,000 acres) of Jakarta, in Indonesia, are used to grow food; and Shanghai is not far behind: more than 10,000 hectares (25,000 acres) of that super-modern city are used to grow vegetables.

These examples are in 'developing' countries, it's true – but the North is catching up. In Cleveland, Ohio, for example, two university researchers set out to determine if their city of 400,000 hungry souls could feasibly achieve self-reliance in the provision of several key foods. In a street-by-street survey, the researchers first identified more than 18,000 vacant lots that could potentially be put to work and then, with a focus on foods suited to urban production (vegetables, fruits, chickens, and honey), they calculated likely yields. Their conclusion: if three-quarters of Cleveland's currently available vacant land were to be utilized – together with a modest number of industrial and commercial rooftops – then the city could provide nearly all the vegetables it needs, more than 90 per cent of poultry and eggs, and all of its honey.[21]

These positive results in Cleveland were confirmed by a project in England that I was involved with. In Middlesbrough – like Cleveland, a rust-belt city intent on self-renewal – I commissioned a large urban agriculture project as part of a social innovation biennial called Designs of the Time. As with Cleveland, we first identified locations for productive growing across town, and then helped community groups, volunteer organizations, school students, public health workers, and even preschoolers to grow food on these sites. Locations ranged from school playgrounds and hospital car parks, to seedbeds outside a hairdresser, the lawn outside an art gallery, and a university campus. It took little encouragement to persuade more than one thousand people in sixty community groups to take part. As their different crops ripened – butternut squash, tomatoes, and other produce – our urban farmers took them to meal assembly centres, in different locations, where they learned how to cook what they had grown. The project culminated in a celebratory 'Meal for Middlesbrough' in the city centre; seven thousand people came to lunch.[22]

A NEW METABOLISM

One lunch, I concede, will not transform a dysfunctional food system on its own – but the social energy unleashed in Middlesbrough is typical of urban food projects around the world. I think of these projects as seeds, landing on harsh-looking ground, that unexpectedly flourish and transform a city's metabolism. The sheer variety of seedlings is a marvel to behold. People across London, for example, are growing hops in a mosaic of back gardens, patios, balconies, allotments, and community gardens for use by local brewers in the Brixton Beer project. This patchwork farm has practical advantages over industrial hop-growing: aphids and mildew can sweep through hops on a large plantation with devastating results, whereas in a more dispersed context, and a more biodiverse setting, this risk is reduced.[23]

Does urban resilience mean a brewery in every basement? In the US, where the number of new craft breweries is at a 125-year high, the development of so-called green breweries has become a driver on its own account of economic revitalization. When the craft-beer impresario Jimmy Carbone pondered running for mayor in his childhood hometown, Haverhill, Massachusetts, he started a project called Shoe Town to Brew Town. Could an old 'shoe city' like Haverhill be transformed into a resilient green 'brew city', he asked – one that would also be a demonstration site for industrial and biological symbiosis?[24] Carbone's quest led him to investigate possible new uses for industrial buildings that had fallen into disuse, and he asked his friends in craft brewing, green building, engineering, and microbial science for advice. Unlike energy-intensive and water-guzzling industrial brewing, he discovered, craft breweries can use their waste products as raw materials in other productive processes in a local ecology; the New Belgium Brewing Company in Colorado, as an example, uses 40 per cent less energy per barrel of output than the average American brewer because – from hops in, to beer out – every stage of the firm's brewing process has been designed for greater efficiency and the reuse of waste.[25] Brewing wastes can also enrich food chains; the mash left over from the fermenting process, which is microbially rich, can be fed to pigs, fish such as perch, or oysters and mussels. In Chicago, food pioneers are turning a former meat-packing plant into a vertical urban farm that combines aquaponics – a super-efficient plant-and-fish-growing system – with kombucha tea production, beer brewing, biogas energy, and a kitchen that serves up the end result with net-zero waste.[26]

If beer is one essential of urban food, bread is another – and the Real Bread Campaign, in the UK, also combines decentralized small-plot growing with local production. As a substitute for additive-filled industrial bread that's transported around the country in diesel-guzzling trucks, the Real Bread people are creating shorter 'grain chains' that minimize the distance between where grain is grown, the flour milled, the dough baked, and the bread consumed. One pilot, the Bedale Community Bakery, uses flour that has been milled just 3 km (2 miles) away at a restored water mill. In London itself, Brockwell Bake grows heritage wheat on allotments, in school and community gardens, and with farmers close to the city; network members have access to a stone mill, and bake with heat throughout the year; 'real baking' workshops are run with local schools.[27] The campaign is innovating new business models, too. In a growing number of Community Supported Baking schemes, the risks and rewards of each enterprise are shared; the Handmade Bakery in Yorkshire, for example, is a worker-owned cooperative.[28] A real-bread movement is also flourishing in Bulgaria, where a Bread Houses Network operates centres in twenty cities. People of all ages and backgrounds gather in them to prepare bread around candle-lit tables and bake the loaves in traditional fire ovens. A Bread Route has been launched that links Bulgaria's Bread Houses to similar centres in Italy and countries in between. A global network, Bakers Without Borders, has also been established that helps communities develop similar bread ecosystems in other parts of the world.[29] A project in the US called Mobile Bread House has similar ambitions. A community bakery on wheels is equipped with a wood-fired oven, a big round table where people make bread and meet each other, shelves full of herbs and teas from various countries, and a hydroponic green roof on which grows wheatgrass from different countries as well as fresh herbs and other plants.[30] As these lattice-works of activity, infrastructure, and skills expand, regional-scale 'beer sheds' and 'grain sheds' are beginning to take shape.

COLLABORATIVE DISTRIBUTION

These hybrid urban-rural models – grainsheds, fibresheds, foodsheds – are not an immediate replacement for the whole of industrial agriculture, but they are proving to be thriving components of a multi-dimensional approach that is turning the global system on its head. Rather than drive the land to yield more food per acre, production is determined by the capacities of land, both inside and outside the city. 'Growth', in this new food ecology, means land, soil, and water getting healthier, and communities more resilient. It's also a system in which everyone, not just distant farmers, has an active part to play.

A next-generation community-supported agriculture platform called La Ruche Qui Dit Oui ('The Hive that Says Yes') is a harbinger of this new approach.[31] Conceived in France, but spreading rapidly to other countries as The Food Assembly, La Ruche combines the power of the internet with the energy of social networks to bridge the gap that now separates small-scale food producers from their customers. La Ruche is the brainchild of Guilhem Chéron, an industrial designer and chef, who drew on expertise gained during fifteen years in the design industry to develop a way for people to connect directly with the farmers who grow their food. La Ruche replaces the long chains of intermediaries in traditional food systems with local 'hives'. When someone starts a local hive, she recruits neighbours, friends, and family to join – the ideal number seems to be thirty to fifty members – and the group then looks for local food producers to work with. These farmers offer their products online at a desired price, and hive members pay 20 per cent on top of that to cover a fee for the hive coordinator, a service fee to La Ruche, plus taxes and banking costs. La Ruche makes money as a platform provider, but it is not an intermediary: the farmer receives the price asked for, and the system is fully transparent; everyone involved knows what happens to every cent transacted. Support teams at La Ruche look constantly for ways to make the system more adaptive; for example, if a whole cheese is too big for a single family, the system enables quarters and halves to be supplied in a way that works for both sides of the exchange. New tools are also innovated; one such is an app that links the scales used to weigh each order to an iPhone. This eliminates paperwork and streamlines order processing.

The social element is integral to the Food Assembly model; weekly face-to face contact between customers and farmers is a core feature. Delivery and distribution of orders happens one day per week at a convenient location; producers meet their customers when they hand over the ordered produce. Members usually make time for a drink and a chat. 'I didn't invent anything new,' Guilhem told me, 'I just merged different ideas into a service that worked better for the people involved. We've designed a tool that can make the distribution and exchange process ten times faster, and easier, and be more appealing than existing systems.' An important aspect has been the creation of a social enterprise model that supports the livelihoods of farmers and hives; it's therefore more resilient than a voluntary organization.

SOCIAL FARMING

Learning about La Ruche triggered a question that had been
bothering me for ages: why do the rest of us regard it as normal
to do literally none of the work involved in growing the food upon
which all our lives depend? It cannot be healthy that we leave that
task to remote farmers who we do not know, and therefore do not
help. A growing number of city people, having come to the same
conclusion, are getting involved in new and diverse kinds of farm-
based activities. In so-called care farming, for example, the assets
of a farm begin to serve the needs of ageing populations – not just
by feeding them, but in animal-assisted therapy and therapeutic
horticulture. On a growing number of learning farms, too, city people
are picking up skills in thatching, dry-stone walling, hedge laying,
hurdle making, and charcoal making, as well as growing food.
The range of experiences available as agritourism is growing. And
many redundant farm buildings are finding alternative uses as a base
for rural social enterprises started by city people. These and other
forms of diversification bring practical benefits, such as new cash and
employment for the farmer and her local economy – but the more
important outcome, for me, is the proliferation of social connections
between city and country. These connections begin to cure the
metabolic rift that I wrote about in Chapter 1.

There's a vast amount for city people to do – even if, at first, it's
part-time. For one thing, ecological agriculture is far more knowledge-
intensive than the industrial model it's replacing – and multiple
skills are needed to cope with that complexity. Each farm has to be
understood and designed as an ecosystem within a bioregional web
of other natural systems. The carrying capacity of the land must be
determined, and systems put in place to monitor progress and feed
back results. It means growing different crops and animals in synergy –
as far as possible, organically. It requires open information channels for
the sharing of resources. It means the development of new distribution
models, of the kind discussed above.

Few of these needed innovations lend themselves to being
invented in a lab or university. What's needed is a knowledge

ecosystem that's rooted in its place – and successful examples to learn from already exist. At the Shashe Endogenous Development Training Centre in Zimbabwe, for example, where huge cattle ranches were once owned by three farmers, 365 families now manage the land sustainably and have proved, in so doing, that it is possible to learn about ecological agriculture in practice.[32] This living example of small-scale but high-quality ecological farming is celebrated in the so-called Shashe Declaration of 2011. This historic document pledged support to small farmers the world over for their expertise as stewards and managers of diverse agro-ecological systems. It acknowledged their capability to integrate farm, forest, and trees in complex, diverse, and dynamic mosaics. It praised their skills in water conservation and harvesting, watershed management, agroforestry, and their protection of the genetic diversity of seeds. The Shashe Declaration acknowledged, too, that the world has much to learn from the ways the farmers in Zimbabwe combine traditional and contemporary knowledge – and share that knowledge with each other, and with other bioregions. Shashe's farmers have developed new ways to share resources and infrastructure – such as storage facilities, roads, electricity, and information platforms. They are also developing new forms of credit and insurance to guard against weather-related risks. The sharing culture found in Shashe by La Via Campesina is being replicated; the network runs regional training programmes, organizes exchange visits, produces educational materials, and distributes success stories between bioregions, on a modest global scale. Its next step is to extend this network of ecological agriculture training programmes, and to use farmer-to-farmer and community-to-community networks to recruit participants.[33]

The not-so-secret ingredient in a healthy but complex local food ecology is the embodied presence of those involved. As we saw with Operation Hope in Chapter 2, ecological agriculture cannot be practised remotely. The eminent ecologist Fred Kirschenmann confirms this important finding: the only way to manage any landscape sustainably is by living in it long enough, and intimately enough, to learn how to manage it well. 'Nature is complex, and constantly evolving,' he explains; 'if one does not live in close conversation with nature, mistakes will be made that are harmful both to one's self and one's place.' Farms, ultimately, are biological organisms, Kirschenmann has concluded; this principle holds true for managing agricultural landscapes as well as wild lands.[34]

Although small-scale farmers appreciate direct support such as seeds, fertilizers, and cash, they value indirect and longer-term support even more – and one way governments have begun to make a practical long-term difference is through public procurement. In the UK, for example, where more than £2 billion of public money is spent each year on meals, a growing number of hospitals, schools, nurseries, care homes, prisons, military units, and central government departments are purchasing ecological food at fair prices from smallholders.[35] The pattern is evident across Europe; government agencies are connecting the dots between direct support for small-scale farmers and the provision of healthy food to children, sick people, and government employees.[36]

At a higher level, too, there are many signs that governmental policy is shifting focus towards the small farm sector. Politicians have finally understood that because small farms grow most of the world's food, it makes no sense that industrial agriculture should hog 80 per cent of subsidies and 90 per cent of research funds. Brazil, one of the world's largest agricultural economies, has launched a National Plan for Agroecology and Organic Production.[37] France and Switzerland have included agroecology in their agricultural policy frameworks. And an EU-wide project called Farm Path is supporting the diversification of farm-based activities and new forms of land use, in the interests of increased biodiversity.[38] Multinational agroecological movements, which have been active in Latin America for thirty years, are now growing in Africa, too; a new Food Sovereignty movement is promoting an agro-ecological approach.[39] And in Mali, in 2015, an International Forum on Agroecology brought together peasants, family farmers, fisher folk, pastoralists, indigenous peoples, agricultural workers, consumers, urban poor organizations, NGOs, academics, and other social movements. The meeting launched a programme to exchange local knowledge and knowhow of farmers, share the innovations of peasants and small-scale food producers, and strengthen synergies between the different small-scale food producer organizations, social movements, and other organizations promoting agroecology.[40] The director-general of the Food and Agriculture Organization of the United Nations (FAO), until recently perceived to be a sworn enemy of the small farm movement, celebrated 'the opening of a new window in the cathedral of established thinking'.[41]

The food-system picture is fiendishly complicated – and this is a healthy sign. We live at a time when ecocidal and healthy approaches to food coexist. Production agriculture is rampant in some parts of the world – but a million seeds of an alternative way are also thriving, even under pressure. I've told you about a few of them in this chapter. These hopeful stories are worth celebrating for their own sake, but I also take hope from the cases in which such seeds have escaped from their tiny niches and occupied the whole landscape. One of those examples is Switzerland. Starting with a small group of organic farmers in the 1930s, a long slow process of change accelerated dramatically in the 1980s when two crises shook Swiss public opinion. The Chernobyl disaster, and an accident at Sandoz, which released agrochemicals into the Rhine, acted as tipping points: they persuaded the Swiss public that something more radical was needed than a 'do less harm' approach. This change in attitudes led to the implementation of a new agricultural law, in 1993, which made ecological agriculture – with its whole-systems focus on low-inputs and local knowledge – Switzerland's national policy. Switzerland is not a typical country, of course – but then, which country is?[42]

FOOD AS A COMMONS

Running through the stories in this chapter is a green thread: the efforts of people in diverse contexts to reconnect to their food – where it is grown, by whom, and under what conditions. These practical, local, and human-scaled activities are the seedlings of an alternative to an industrial food system that, as an extractive industry, is as cruel to people as it is to animals, and the land. The right approach is well enough understood: plough the soil as little as possible; keep the soil covered; increase biodiversity. These are more than instructions – they are values. The new food systems are as much social as they are technical experiments; they're happening wherever people organize together in new ways – not just to grow or obtain food, but also how to live on and steward the land. Community land trusts, for example, are a welcome innovation in the legal basis of social farming. The Food Commons in Fresno has just such a plan; so, too, does the Fordhall Community Land Initiative,[43] in England, where one small farm now has eight thousand landlords. This commons-based approach to food is about cooperation, sharing, and stewardship.[44]

6 CLOTHING: FROM DIRT TO SHIRT, AND SOIL TO SKIN

I arrived, as one does, at a lingerie factory in Sri Lanka. It boasted impressive rainwater capture tanks, anaerobic digesters, and a water-saving sculpted landscape. There is fierce competition between different fashion manufacturers in Sri Lanka to prove that their factory is greener than its competitors More than one million people (out of a population of 20 million) work in Sri Lanka's fashion industry, so it's critical to the whole economy. Companies have to contend with two pressures: on one side are powerful foreign buyers such as Marks & Spencer (the sole client of the factory I was visiting), Tesco, and Victoria's Secret; the latter's buyers tend to arrive in large helicopters to inspect their mostly tiny products. Sri Lanka's industry must also contend with competition from other fashion-producing countries, from Turkey to Bangladesh, which also support hundreds of thousands of small and micro businesses. The big global buyers can and do switch production from one country to another at a moment's notice if they deem it necessary to do so for competitive reasons.

Squeezed like this, it's no small matter that Sri Lanka has resolved to compete on the basis that its companies are ethical and sustainable, not just cheap. A centrepiece of this strategy is a platform, called Garments Without Guilt,[1] that commits the country's apparel firms to protect workers' rights, create opportunities for education and personal growth, and help alleviate poverty in local communities. At the symposium on ethical fashion that was my reason for the visit, we heard a lot about the buyers for huge brands whose only interest seems to be to drive down prices. We heard how fast fashion seems to be accelerating of its own accord, with newer products being launched at ever shorter intervals. We heard that young people seem to be unconcerned by big-picture issues, and drive this manic consumption along. I learned about a special problem: if you launch a line of 'ethical' products, they make all your other products look, by implication, unethical. One industry leader told me of his pride that Sri Lanka has evolved from a 'nation of tailors' to a 'master of integrated solutions'. This did not sound to me like progress. I begged him to look after and nurture the country's tailoring traditions – but look for new ways to make them viable as a business. Connect the people who make things, here, I pleaded, with people who need clothes and would love to have a direct relationship with the person who makes them.

At first sight, the fashion industry is indeed ethically challenged; you probably need to be naked to read this paragraph with a clear conscience. I learned from Kate Fletcher's landmark study, *Sustainable Fashion and Textiles*, that it took 2,700 litres (700 US gallons) of fresh water to make my cotton T-shirt – and much of that water ends up saturated with pesticides; a quarter of all the insecticides in the world are used on cotton crops.[2] It's partly down to me that 85 per cent of the Aral Sea in Uzbekistan has disappeared – because its water is used to grow cotton in a desert. Transport costs are also a big issue – the average T-shirt travels the equivalent distance of once round the globe during its production – so now we have to worry about fabric miles as well as food miles. Buckets of hazardous sludge are generated during the coating process used for the metal buttons on my jeans. White is energy-intensive because of all the bleaching. I'll use six times more energy washing my favourite shirt than was needed to make it. Nearly all the textiles in my life will end up in landfill – garments, household textiles, carpets, the lot. Being clean, and wearing white to prove it, has weakened my immune system.

Global food and textile systems are interlinked to a surprising degree. The big discount chain Primark, for example, which is famous for its ultra-low prices and celebrity customers, is owned by Associated British Foods (ABF), a multinational food, ingredients, and retail group. ABF is adamant that its affordable clothing is produced 'efficiently, and to a high quality'. This may well be true, but the problem is that the words 'efficiently' and 'quality' do not, of themselves, mean 'sustainably'. The cheap cotton used by Primark takes just as long to grow, and uses just as many resources per unit of weight, as the cotton in a $200 T-shirt; besides, the prices of its virgin fibres are so low that there is no economic incentive for anyone to collect, sort, distribute, and resell the clothes its millions of customers discard. Over 90 per cent of resources entering the system of discount stores like Primark are discarded as waste within three months. Even if firms like Primark were to use only bamboo and soya bean fibres, grow 100 per cent organically, and produce only locally, their T-shirts would *still* not be sustainable because of what happens when we get a garment home. The average piece of clothing is washed and dried twenty times in its life: 82 per cent of its lifetime energy use, and over half the solid waste, emissions to air, and water effluents it generates, occur during laundering. Among the many quirks in the fashion and textile system is the fact that a $200 T-shirt has a heavier environmental impact than a Primark one – because it gets washed, rather than chucked away.

It is one thing to draw attention to the hidden costs of fashion – quite another to figure out what to do about them. At the Centre for Sustainable Fashion, London, director Dilys Williams told me that many much-trumpeted 'solutions' turn out to be less perfect than we hoped, or were told. So-called biodegradable fibres, for example, cannot be chucked on your compost heap – as I, for one, had assumed; the near-ambient conditions of home compost heaps do not provide the right temperature and humidity. PLA fibres (as some of the biodegradable ones are called) decompose only in the optimum conditions provided by an industrial composting facility – and there are very few of those in the world. Another surprise: natural textiles can be more harmful than synthetic ones. Although polyester fibre, to take one example, is made from non-renewable petroleum and requires large energy inputs to produce, it is not *so* environmentally damaging when its whole life cycle is calculated – from sourcing the raw materials, through the use phase, to the disposal phase. Polyester has lower energy impacts than cotton during the washing and cleaning phase, for example; it is also completely recyclable at the end of its life.

Although attitudes across the industry are changing, the pace of globalization, too, is accelerating. Many international brands now source products from within large industrial zones in poor countries where the identity of individual suppliers is obscured; huge quantities of lethal toxins in wastewater, for example, cannot be traced to individual facilities. In Ethiopia, a vast global hub for the shoe and accessory industry is the workplace of 200,000 guest workers whose wages, at US$35 a month, are ten times lower than those in China. Seven-year tax breaks, regulatory exemptions, cheap land, and 'cost-sharing schemes for foreign experts' are among additional attractions for global firms.[3] Cheap labour is not the only driver of this global reconfiguration of fashion production. Cheap raw materials, and cheap nature, are an added incentive. Herders and family sheep owners are paid $10 each for hides that end up as coats sold in rich countries for $500 upwards. In Ethiopia's case, industrial expansion on this scale portends terrible damage down the line for soils, rivers, and watersheds – and the people and animals who steward them. Ethiopia is among the most water-stressed countries in Africa, for example, and increased tanning, with its prodigious water consumption, exacerbates that; it takes 16,600 litres (4,385 gallons) of water to produce 1 kg (2¼ lbs) of leather.[4] Tanning also has one of the highest levels of toxic intensity, per unit of output, of any industrial process.[5] During the different procedures used for hide preparation, tanning, and finishing, at least 300 kg (660 lbs) of chemicals are added per ton of hides. The huge discharges of air, liquid, and solid waste pollution generated by tanning contain chromium, copper, cadmium, and other toxic by-products. Effluents released on the land, or dumped into the surface water, are associated with skin blisters, diarrhoea, gastroenteritis, urinary tract infections, and liver diseases among workers and their families living near tanneries.[6]

In collecting perplexing examples like the ones above, I've developed both respect and sympathy for fashion industry professionals. Although fashion designers, for example, are commonly portrayed as shallow *artistes*, they must not only master a lot of knowledge about style, cut, fabric, colour, and design; they also have to navigate the technical and business worlds of textile processing, marketing, communications, and distribution. These issues are complex enough. To be told, as is happening now, that their design decisions impact on watercourses, air quality, soil toxicity, and human and ecosystem health, in other parts of the world, is incredibly hard to deal with.

DOING LESS HARM

Thanks to the tireless work of activists and advocates, millions of people are now aware of the social and environmental harm wrought by industrial systems – including the textiles and fashion ones. As awareness and concern has grown, many fashion brands have committed to do less harm and proclaim that their products are verified, accredited, or certified as being sustainable. But the barrier to adherence has been low; in the absence of a shared definition of what 'sustainable' actually means, more than 400 different textile and clothing industry sustainability labels and standards commit brands merely to 'minimize' negative economic, environmental, and social impacts and do 'as little harm as possible'.[7]

Some big firms, it is true, are now moving beyond that empty promise. The Sustainable Apparel Coalition, whose commercial members represent more than a third of the global apparel and footwear industry, has launched an assessment tool – the so-called Higgs Index – that standardizes the way social and environmental impacts are measured, from materials to end of life. Energy use, waste, and toxic outputs from materials, manufacturing, and packaging will all be considered. But even this addresses the symptoms, but not the principal cause, of our difficulties: an economy based on perpetual growth in a finite world. Not a single fashion brand has told its customers to buy less.[8]

KEEP YOUR STUFF ALIVE

Change leaders in the worlds of fibre and fashion have learned the hard way that exhortations to 'buy less, wash less' are ineffective, on their own, against a global 'Two Wash, Two Wear' industry; the financial DNA of the system as a whole compels it to grow at all costs. The new approach is to seek to influence the fashion system at its edges, using cultural interventions as a tool.

At the London College of Fashion in London, at an event I attended called 'Craft of Use', 200 professionals explored the question: what kind of system would improve the quality of our fashion experience without increasing the quantity we consume? For thousands of years before the oil age, textiles were carefully looked after; the repair, alteration, and maintenance of clothes was a normal part of daily life. Can we not combine the beauty of that culture with peer-to-peer production? The main input to our discussions was a remarkable archive of 500 stories, collected by designers and artists around the world, in which a sustained attention to wearing, tending, and caring for clothes was a source of satisfaction and meaning.[9] In her project Grasslands, for example, Emma Lynas connects garments to the earth literally, by extracting colour from her agricultural grassland in Victoria, Australia. She collects eucalyptus leaves, wild thyme, cedar berries, and Aleppo pine needles – and simmers them with strips of hemp. The result is a range of summer hues: straw, gold, and bronze. In another example, Sasha Duerr forages for materials in her neighbourhood. She uses plants directly, rather than extracts. She uses contextual knowledge about the life cycles of plants, their seasonal availability, and their colour potential to plan commissions – much as an organic chef plans menus around locally and seasonally available food. (For her wedding, Duerr hand-dyed all of her bridesmaids' dresses using fennel that she gathered from around her neighbourhood.) Together with the University of

California Botanical Garden in Berkeley,[10] Duerr has now developed a Fiber and Dye Map that details the plant names, their uses and properties, and the colours and textures that their leaves, fruit, and bark are capable of producing.[11] The mapping project led, in turn, to the creation of a brand new plot dedicated to edible fibre and dye plants at the Berkeley botanical garden. 'Recovering knowledge can be a creative process,' says Duerr. 'I love rediscovering and experimenting with "lost" recipes. We have to raise awareness of how rich our world is, of how many possibilities there are. You go to the store and there are only three or four different kinds of apples to choose from. With fibre and dye plants it's the
same way. There is so much more out there for us to explore!'

A Dutch designer, Christien Meindertsma, in her Flax Project, also explores the life of the products and raw materials that have become so invisible in an increasingly globalized world. Meindertsma makes a series of products from flax produced locally – from the seed to the end product. Documenting the production process is an important part of the project; Meindertsma collaborates with a film-maker and a photographer to make quite beautiful records of the five stages of flax processing: sowing, blossoming, harvesting, retting, and pressing bales.[12]

A key learning from these stories is that people who care for their clothes through time do not regard them as inert, static objects; they do so, in Kate Fletcher's words, in 'a lifeworld that is itself a source of meaning'. For Fletcher, who conceived 'Craft of Use', people who connect with their clothes are more likely to connect with their makers, too, and with the ecosystems from which their materials come. An awareness of natural systems – of cycles, flows, webs, and interconnectedness – comes naturally to people who are close to the whole life cycle of their clothes. For Sabrina Mahfouz, our poet in residence for the day, the lesson was simple: reconnecting with the neglected qualities of material is a key to system transformation: we must 'Keep Our Stuff Alive'.

FROM DIRT TO SHIRT, FROM SOIL TO SKIN

In food systems at the scale of a bioregion, as we saw with The Food Commons in Chapter 5, transformative change is growing from a grassroots level. So, too, in fibre. In California, the Sustainable Cotton Project (SCP) is changing the ways that cotton, one of the world's most widely traded commodities, is farmed and marketed. As Lynda Grose explained it to me, its approach reduces the toll that the soil-to-shirt cotton production process takes on the Earth's air, water, and soil, and on the health of people in cotton-growing areas. SCP helps small, and medium-sized conventional cotton farmers produce a high-quality fibre while minimizing its chemical impact; by reducing chemical use, and integrating more biologically based management practices, these steps help farmers become better stewards of the land, air, and water in their bioregion. Grose encourages growers to look at their farms as a whole system, and helps them use biological controls, rather than chemical ones, as their first line of defence. SCP helps farmers combat pests by providing habitat for beneficial insects; if more good insects are needed to combat a pest, it augments the farm's defences with releases of insects when needed, while providing free weekly monitoring of pest and beneficial insect populations. It's a social as well as an ecological system approach; as each farmer's expertise grows, she becomes a mentor to neighbours. SCP organizes Cotton Farm Tours through the San Joaquin Valley to give interested farmers a behind-the-scenes look at how the 'cleaner cotton' farmers are doing it; they visit perennial hedgerows, walk through coloured cotton fields, and tour a cotton gin. Regular field days and mentor programmes enable further information sharing and mutual support among growers.[13]

Rebecca Burgess, founder of Fibershed in California, is also confident that fibre will follow food in public awareness.[14] She began Fibershed with a challenge, to herself, to wear clothes for a year sourced and dyed within a 240-km (150-mile) radius from her front door. The essential elements for a bioregional fibreshed were in place, Burgess discovered: animals, plants and people, skills, spinning wheels, knitting needles, floor looms.[15] But there was a lack of connectivity between these different actors and technical components. The many small farmers and producers within her region were doing great work – but on a small scale and, for the most part, below the radar. 'Our priority is to integrate vertically,' explains Burgess, 'from soil to skin.' As a first step in connecting the fibreshed's actors, an annual Wool and Fine Fiber Symposium was inaugurated to bring together the region's producers, shearers, artisans, designers, knitters, fibre entrepreneurs, and clothes-wearing citizens.[16] They discuss what is needed next to bring 'farm-fresh' clothing to the region. All manner of fine-grain issues emerge: flock health; rotational grazing; weed management; predator issues; breeding for fibre, colour attributes; milling and fibre-processing capacity. A constantly evolving database collects data on everyone operating a dairy, ranch, farm, or homestead with one or more fibre-producing animals.[17] Data from this Wool Inventory Map is now used to assess the scale, scope, and location of future fibre-processing facilities. Also in development is a prototyping and education facility, called FiberLab.

LEATHERSHEDS, TOO

In the world of leather, too, new models are emerging – and in unlikely contexts. One of these is Kanpur, a tough industrial city on the banks of the Ganges River that is India's largest leather hub. The city has long had a bad reputation thanks to the high levels of pollution that emanate from its many tanning factories. But the leather industry provides jobs and livelihoods for over 250,000 people, and for Mansi Gupta, a designer whose family have leather factories in the city, these workers have to be the starting point for any solution. 'Clean production is at heart a social process, not just a technical one,' she told me; 'my job is to change relationships among the people who make, use, and look after things, and the materials they use to do so.' Rather than begin with yet another diatribe about the evils of tanning, Gupta's first step was to celebrate the leather ecosystem of Kanpur – its people, skills, and cultures – in a publication and short film. Her next step was a series of workshops with change-minded people and institutions in Kanpur's leather ecosystem.

It emerged that clean forms of tanning are being developed in a few small centres in Europe and the US. Leather making has not always been a toxic activity; in pre-industrial times people used natural materials and processes to make soft and durable leather a bit like chamois leather. As agriculture developed over the centuries, the use of bark tannage yielded the firmer, denser leathers used in horse gear, hard-soled footwear, belts, and armour. By the eighteenth and nineteenth centuries, virtually every US town had a tannery to service the local population's need for footwear, horse gear, gloves, and the like. Leather products were easy to maintain and repair locally.[18] In ecologically balanced leather making today, natural tanning agents – tannins, alum, earth minerals, fish oils, and wood smoke – have potential in a modern manufacturing context; the same machinery employed by chemical tanning can be used. As is the case with Fibershed products, costs are somewhat higher. This is in part because the harsh chemicals used in production tanning are occasionally replaced by hand work – but time is another factor. Industrial tanning chemicals, such as sodium sulphide, cut the time required to soak and de-hair hides from about a week, to mere hours – but they do so with extreme force. Hair is turned into pulp in ways that allow the entire de-hairing process to take place in a large spinning drum without human interference. Further time is saved because the hides stay in the drum throughout the wet stage of the tanning process while different chemicals are added and drained out as required – but the downside is a flow of toxic, chemical-laden pulp that too often ends up in the sewage system, where it binds with oxygen, leaving none for the aquatic life.[19]

Although encouraged that alternatives to toxic leather processes exist, Gupta has realized that new business models are needed for these new approaches to flourish. She therefore opened a lab that brought together leather workers, tanning experts, designers, and businesspeople to create prototypes of promising business ideas. Right now, these actions remain small – but, taken together, they begin to tell a new story about the leather industry and the people who work in it. For the first time, leather workers have a name, a face, and a presence at the centre of the story.

The London-based Leather Working Group, whose members include Mulberry, Burberry, Louis Vuitton, H&M, and Dr Martens, requires that the farms and tanners they buy from meet specific environmental stewardship standards, and procedures are in place to ensure compliance.[20] But therein lies the fundamental problem: everyone involved – herders, tanners, producers, intermediaries, brands, and customers – is related to each other in transactions. Each actor along the chain has an incentive to buy low and sell high. Time is money, and nobody wants to pay for it. The system is steered by an army of product managers in bigger companies whose pay and status is determined by profit margins, not by the health of some distant watershed. This commodity-based business model guarantees that the interests of ecosystems, and the mostly poor people who steward them, come last.

A solution is not hard to describe. If one of the big fashion conglomerates, such as Kering, were so minded, it would require its managers to optimize a different kind of margin: instead of the margin being between two abstract numbers – cost and sale – the company would measure the difference between the health of the land before and after each turn of the business cycle. Operating at a bioregional scale, such a big group could surely find ways for the lands where its materials originate to be put in some kind of trust, of the kind pioneered by The Food Commons in the USA; this trust would lease the land to small herders subject to various mutually agreed conditions. Each bioregion would be governed collaboratively to the same end: the regeneration of its living systems. Kering would still get its materials, but its products would have a higher value. By helping its customers feel connected with everyone involved in an item's making, and the unique skills and resources of a particular bioregion, the group's brands would transcend the impersonal and anonymous transactions associated with the industrial model.

LAND BEFORE 'ECONOMY'
What will it take for the new to replace the old – and not just coexist with it? Fashion is a supremely adaptive system. Its participants –

including designers – pay super-close attention to minor changes in their market as a matter of course. Until now, their attention has not extended to the social and ecological health of the bioregions their materials come from. The next step, surely, is for the global ecosystem of sustainable fashion pioneers to put the health of the land, and the people who live on it, at the centre of their story.

Alternatives to the global commodity food and fibre systems are now emerging in myriad diverse experiments. Ranging from farmers using non-genetically modified cottonseed, to artisan makers who access markets and coordinate supply webs using web-based platforms, these are the harbinger of a 'leave things better' and regenerative economy. As Lynda Grose describes it, these experiments turn the global system on its head. Rather than constantly driving the land to yield more fibre per acre, production is determined by the land's health and carrying capacity, which is constantly monitored. Decisions are made by the people who work the land and know it best. Fibre prices are based on yields the land can bear, and on revenues that assure security for the farmer. 'Growth' is measured in terms of land, soil, and water getting healthier, and communities more resilient. 'The commons approach provides a way out of the polarized squabbling between companies and activists,' says Grose. 'It removes the need for complex – and for the most part misleading – accreditation and labelling systems.' Social patterns of organization coevolve with this diversification of material flows. Some regions form cooperatives, others farmer/worker-owned entities; some function as food-and-fibre commons. These community forms of social solidarity pave the way for new terrains for fashion practice. And citizens, instead of being treated as passive consumers whose task is to speed the flow of materials through the fashion system, are recast as collaborators.

FIBRE NETWORKS

A remarkable example of the fibre community's propensity to collaborate is the website Ravelry, an online forum of four million (at the time of writing) knitters, crocheters, designers, spinners, weavers, and dyers. Launched as a free social networking service by husband and wife Casey and Jessica Forbes, in 2007, the site enables this vast community to keep track of their yarn, tools, and project and pattern information, and look to each other for ideas and inspiration. In addition to serving as a social network site for users to discuss their craft, Ravelry facilitates micro-business, and allows crafters to sell their patterns or yarns. For Dr Sal Humphreys at the University of Adelaide, a key to Ravelry's success is the skilful way it harnesses the capacity of the internet to handle micro-transactions and cater to niche markets, while remaining a trusted hub for the social ecosystem of the fibre communities.[21] 'It is not an unregulated free for all,' says Humphreys, 'there is a hierarchy of decision-making and interventions available. But there is also the pervasive notion of "the spirit of Ravelry" at work within this corner of the fibre economy.' A culture of gifting is an important part of this 'spirit': many people knit items that they give away to friends and family, and this dovetails with the rhetoric of generosity that seems to be a key to the site's success.

The complexity of activities and relationships supported by Ravelry, and the variety of formats used to support them, would be impossible on a purely commercial site. Its outputs are sometimes social rather than material; timeframes for production are highly varied; and the benefits are sometimes financial, sometimes social, and sometimes both.

Until now, because we've lacked a shared definition of what 'sustainable' actually means, we've been fobbed off with empty promises to leave the world 'as unspoilt as possible'. At the scale of a fibreshed, textile production can, for the first time, be monitored against the health of soils and watersheds. Regional-scale production cannot supply as many cheap clothes as the global system does – but is an ultra-cheap T-shirt a more important 'need' than keeping the land healthy? I don't think so. Besides, for the many millions of people already active in maker networks like Ravelry, a regional fashion ecosystem, and knowing everyone involved in a garment's making, can transcend the impersonal and anonymous transactions associated with the industrial model. Keeping our stuff alive means keeping the land alive.

7 MOVING: FROM TWO-WHEELED FREIGHT, TO CLOUD COMMUTING

The big Audi that collected us from Istanbul airport contained nearly as many electronic control units as the new Airbus A380. The Audi, and similar high-end cars, will soon run on 200 million lines or more of software code.[1] As a comparison, the avionics and onboard support systems of Boeing's new 787 Dreamliner run on fewer than seven million lines. That makes modern cars highly intelligent, right? Well maybe, and maybe not. Suppose the owner of such a 2-ton vehicle drives a mile down the road to collect a 300 g (10½ oz.) pizza for a small child's dinner: is that a smart thing to do? And if it's not, whose judgment is at fault: the car's, or the driver's? And if locusts that fly in swarms of millions don't bump into each other, how come today's cars 'need' so much computing?

The Audi Urban Future Award, whose jury I had been invited to chair in Istanbul, did not evade this and other tricky questions – for a lot was at stake. Audi's in-house future watchers had noticed an unsettling trend in visions for the future of cities: an increasing number of these visions did not contain cars.[2] Future scenarios for the city seemed to be converging around car-free solutions to the problems of debilitating gridlock, lack of space, and air pollution. Wondering what this trend meant for a car company such as itself, the company launched its Urban Future Initiative to establish a dialogue on the synergy of mobility, architecture, and urban development. A wide variety of experts were asked to explore new ways for cities to think – collectively and holistically – about mobility. The jury that Audi invited to judge them was also diverse: there were architects and mobility experts, of course, but our number also included a writer, a philosopher, a curator, a network designer, a solar-power engineer, and a film-maker.

This was not a concept competition. Each team was required to represent a real city and base its proposal on a deep investigation of each city's context. Most of the entries confronted the core dilemma head-on – cars or no cars? – and some of them pulled no punches. In one distressing video on show in the exhibition, a Chinese man, carrying a baby in his arms, says to camera, 'It's a terrible feeling to walk in the city.' In another film, this one made in Istanbul, a man tries to push a baby along a pavement in the snow; his way is blocked by trash bags on the pavement but he can't go round them because a line of cars is parked nose-to-tail next to him in the road. Considering that 40 per cent of the time we spend travelling, across all cultures, is spent walking or waiting, the conclusion was stark: that the car is complicit in a wildly inequitable use of space.[3]

The response of the Istanbul team was an online loyalty platform called Park. This would harness the power of social networks to increase the use of shared transport. Reduce the presence of parked private cars, their thinking went, and space would be freed up for shared social and cultural activities. The US design team proposed to bundle all systems of transport into a highly technical, optimized, and continually flowing main artery for mobility. The Mumbai-based collective Crit entered a remarkable set of tools, with the family name of Being Nicely Messy, that were designed to enhance a city's 'transactional capacity' – that of fostering valuable connections among diverse populations.[4] A Chinese design firm focused on the radical concept of 'cloud logistics', and proposed to bury the entire transport infrastructure underneath the truck- and pollution-damaged streets of the Pearl River Delta. A joyful celebration of movement for its own sake – in both physical and social ways – was celebrated by Urban Think Tank from São Paulo, in Brazil. In the event, the jury selected Höweler + Yoon Architecture's Shareway as its overall winner because its concept of 'opportunity without ownership' involved both social as well as technical innovation at a system-wide level.[5]

IS MOBILITY A BASIC NEED?

In reflecting afterwards on the Audi Award, my thoughts returned to the small child and her pizza. When she is old enough to drive herself, the ways we occupy our cities today will surely strike her as crazy. She'll be shocked, looking back, by our greedy use of space, matter, energy, and land – just to move around. She'll grieve at the ways our unchecked mobility damaged the biosphere, our only home. The entries in the Audi Award, I concluded, were constrained by two questionable assumptions: first, that mobility is a universal need; and second, that mobility is a technical problem amenable to being solved by engineering means.

The proposition that mobility is a fundamental human need sounds uncontroversial, but think of it this way: one could also say that locusts have a universal need for lunch. Which they do. But when locusts fulfil all *their* needs, the land is stripped bare – and the locusts, having eaten their last lunch, expire. The consensus in archaeology and anthropology is that mobility, far from being an innate feature of human behaviour, is determined by the needs of communities at a particular place and time.[6] Individuals, households, and larger groups move around a lot – or not – depending on patterns of land tenure and their access to land, the capacity of the commons to support them (or not), and other socio-economic factors. Mobility, in other words, is a second-order 'need'. We move *as much as we have to* in order to obtain food, shelter, security, and the opportunity to connect and transact with each other.[7] The more those amenities are present in our immediate surroundings, the less we tend to move. This is why economic localization and sustainability are subtexts of the same story.

This story is a challenge for promoters of 'smart cities' and 'intelligent transportation systems'. Excited by the potential of cars to communicate with us and with each other in amazing new ways, they cite experiments such as Google's auto-piloted cars, or 'smart' sensor-encrusted roadways, as evidence that coordinated communication will soon make car blight a thing of the past. But these promises will end in costly disappointment as long as cars, and their owners, think only about themselves – and *by* themselves. More data for its own sake will not make a city 'smart' if all that computational power is misdirected. On the contrary, it's likely that high-tech complexification will make things worse. Throughout history, each new transport revolution has proved far more expensive to maintain and operate than was anticipated – and the 'smart' schemes being floated now will not be an exception to that rule.

As I wrote in my book *In the Bubble*, our conquest of distance in the modern world has come at a heavy cost. Land, for example, is a finite resource – but we consume it as if it were limitless, especially for mobility. John Whitelegg, a transport ecologist, worked out that in Switzerland the land allocation for road transport is 113 square m (1,200 square ft) per person – and for all other living purposes, such as houses, gardens, and yards, it's 20–25 square m (215–270 square ft) per person.[8] The knowledge economy, far from reducing our greedy consumption of land, accelerates it: the spread of car parking around universities, hospitals, and airports stimulates higher levels of car commuting, demands for more road space, and hence land take, in a perpetual spiral. This misallocation of resources is hardwired into our economic decision-making in which the costs of a major project – its construction and operation – are set against highly questionable benefits. In the UK, for example, a multi-billion high-speed train (HST) project has been justified on the basis that each hour 'saved' by an HST business traveller is worth £54 (74 euros or $83).[9] In the same system, a cyclist's hour is valued at half that amount; saving time for a walker is valued at zero. More than two-thirds of the assumed benefits come from these time 'savings'; they completely disregard the time that people now spend productively on trains with the help of their mobile phones and laptops.[10]

To a car company, replacing the chrome wing mirror on an SUV with one made of carbon fibre is a step towards sustainable transportation. To a radical ecologist, all motorized movement is unsustainable. So when is transportation sustainable, and when is it not? Chris Bradshaw, a transport economist, wants planners and designers to respect what he calls 'the scalar hierarchy'.[11] This is when trips taken most frequently are short enough to be made by walking (even if pulling a small cart), while the next more frequent trips require a bike or bus, and so on. If one adheres to this, Bradshaw points out, then there are so few trips to be made by car that owning one is foolish. This lesson seems to have sunk into young people in a big way. In the USA, new car purchases by 18- to 34-year-olds fell 30 per cent between 2007 and 2012.[12]

If, as seems probable, a 'peak car' tipping point has arrived,[13] we can move on from the futile 'car or no car?' debate that has bedevilled discussions of sustainable mobility for a generation. Since any mode of transportation means pushing through air or water over vast distances that will not shrink, a more interesting question arises: at a time of energy transition, what priorities should inform the investments we can afford to make?

A TALE OF TWO TRAINS

Oslo Airport's bullet train reaches the city centre in nineteen minutes. At 210 kph (130 mph) it is not the world's fastest – some of China's new trains will soon reach nearly twice that speed – but Norway's is surely the most macho to look at. Travelling on Oslo's mean-looking machine for nineteen minutes costs about 20 euros (US$30). In India, by contrast, that same amount of money buys you a 3,500-km (2,175-mile) train ride from Kashmir to Kerala. True, the north–south India trip takes three nights and four days, and the cheap carriages can be crowded – but one is bound to ask: in an age of resource constraints, which country has the most advanced and resilient infrastructure?

My question is not rhetorical. Norway will decide soon whether or not to spend a big chunk of its oil revenue endowment on a nationwide extension of its HST network. In the US, too, HST promoters argue that the country has fallen behind even underdeveloped countries in terms of infrastructure and that building HSTs creates jobs and increases productivity. At the Oslo Architecture Triennale, we discussed whether architects could influence this major infrastructure decision – or must they wait passively until there are stations and bridges to design? My take has that 'high-speed railways, yes or no?' is a second-order question. First-order questions concern the kind of society Norway – or California, or India – aspires to become in the decades ahead. In the case of HSTs, three first-order questions therefore stand out in particular: do the true system-wide costs of an HST network justify the investment? Is it sustainable to spend energy on the compression of space and time? And is it really just empty space out there?

TRUE COST INFRASTRUCTURE

Modern mobility comes with a price – but the price tag is not visible and we travellers don't pay it. It is the biosphere that pays – in the form of impacts and emissions that it absorbs from mobility, but which are not measured and are not charged to travellers. We know this to be true because it's happened before. The development of the US Interstate Highway System is a case in point. Its growth changed fundamental relationships between time, cost, and space. These, in turn, enabled forms of economic development that have proved devastating to the biosphere and to society. It is probably true that HST would facilitate another wave of productivity-driven economic growth – but if that pattern of growth is ecocidal, is it the right path to follow?

At face value, the argument for a Norwegian HST network is strong. The short route between Bergen in western Norway and Oslo, for example, is one of the most travelled in Europe. Two flights an hour in each direction carry over 5 million passengers a year – and the country's population is just 4.7 million. The existing Oslo–Bergen train takes 6.5 hours, whereas an HST service would link the two cities in 2.5 hours.[14] But when so-called true cost economics are applied to HST, the proposition that high-speed trains are environmentally far more friendly than cars and aircraft loses credibility. When researchers at Martin Luther University studied the construction, use, and disposal of Germany's high-speed rail infrastructure, they found that 48 kg (105 lbs) of solid primary resources is needed for one passenger to travel 100 km (62 miles). A Swiss study of the carbon footprint of high-speed railway infrastructure concluded that one passenger kilometre on Europe's high-speed rail network is linked with 6.3 g CO_2 from the traffic infrastructure alone.[15] A research group, this one at UC Berkeley,[16] has also measured the vast amounts of environmentally intensive materials that are needed to build such a system. The Berkeley team analysed hundreds of life-cycle processes – from construction equipment (for example, emissions from bulldozers, dump trucks, excavators, and frontloaders) to the supply-chain effects of producing the concrete and steel needed to construct hundreds of miles of track and stations. These immense resource flows were confirmed by the Swiss research consortium. Reporting at the end of 2009, the UIC group assessed more than forty modules of the rail track system: tunnel, viaducts, bridges, the track itself, energy, and signalization equipment. Prices really soar when an HST requires bridges, tunnels, and winding mountain routes to cover difficult terrain. On the flat run from Madrid to Seville, the bridge-and-tunnel share is only 3.8 per cent – but on the line between Würzburg and Hanover, the share is 37 per cent. In Norway, with its mountainous topography, the resource costs and carbon footprint of its tracks would surely be astronomical.

But an HST system is more than the sum of its tracks. Among the other resource-intensive system footprints that necessarily accompany an HST line are:

Space: land is a finite resource, but we consume it as if it were limitless – especially for mobility. Space has to be consumed in large quantities to provide the infrastructure for high-speed travel – just as it does for new motorways and airports.

Energy supply: even if high-speed travel were not a climate change or social problem, high-entropy transport systems depend on finite energy sources. Whether oil and gas are at a peak, or on a plateau, can be debated – but they are finite and no commercially viable renewable alternative offers the same volume and performance. How resilient is that?

First mile/last mile: before a passenger boards a train, she has to get to the station using other means of transportation – the so-called first-mile element. And HST stations are rarely the end-point of her trip; more infrastructure is needed to complete the 'last mile'.

Station and parking infrastructure: many HST stations are multi-modal hubs entailing complex and energy-intensive walkways, doors, escalators, lifts and the like to connect with local public transport and parking lots. HST terminals and stations also contain shopping malls, restaurants, and other service centres not linked with the core service of transporting passengers.

Security costs: enormous and growing material and human resources must be deployed to reduce the vulnerabilities of these complex systems to malfunctioning or attack.

SPACE-TIME COMPRESSION

Although time savings provide the principal economic justification for HST schemes, the expansion of these networks does not, in the long run, give us more free time. On the contrary: we spend the same amount of time travelling today as we did fifty years ago – but we use that time to travel longer distances. The fundamental problem with the HST is not that it burns too much of the wrong kind of fuel. The problem – as with the interstate highway systems that came before – is that it perpetuates patterns of land use, transport intensity, and the separation of functions in space and time that render the whole way we live unsupportable. Something similar has happened in Norway before. When oil was first discovered in 1969, it spawned a generation of sprawling developments. The suburbanization of Jæren swallowed numberless small rural conglomerations. 'Oilville' now stretches more or less continuously from Stavanger in the north to Egersund in the south.

Are there new ways to think about the space-time geography of countries with lots of space? Might we reimagine wide spaces and long distances as assets rather than as obstacles to be overcome? Space, like oil, is a finite resource. Worldwide, space is at a premium. If a country such as Norway has lots of space, doesn't this make it rich? Why try to compress this valuable national resource? Why try to make it smaller?

Maximum dispersal is the settlement pattern of the natural state of nature. As Stone Age economist Marshall Sahlins has pointed out, dispersal is the best protector of persons and possessions in terms of minimizing conflict over resources, goods, and women.[17] In Oslo, we discussed whether to think of Norway as a mosaic of semi-autonomous zones. Could so-called peripheral cities be reconceived as new centres in their own right? By re-examining what makes the regions of Norway distinctive, could new forms of value be discovered as the basis for establishing settlements?

NOT JUST EMPTY SPACE

Shortly after the Oslo Triennale ended, an International Commission on Land Use Change and Ecosystems published a framework for the valuation of undeveloped land, woodland, rivers, and marshes. Researchers had concluded that the global economy was losing more money from the disappearance of forests alone – US$2–5 trillion per year – than through the banking crisis. (The figure came from adding the value of the various services that forests perform, such as providing clean water and absorbing carbon dioxide.) That study, and others like it, placed a question mark on the assumption that the world is full of empty space that we should aspire to fill, at will, with things like HSTs. As we learn that 'empty space' is not empty, it follows that many supposedly clean transport or energy systems are not inherently clean at all – but only somewhat less dirty than the fossil-fuelled systems they are purported to replace.

In the Nordic countries, I learned, some pioneering designers are already sensitive to the hidden value in 'empty' land. Knut Erik Dahl, for example, told me about a remarkable landscape study by Peder Agger and Jesper Brandt titled 'The Dynamics of Small Biotopes in Danish Agricultural Landscapes' (1988).[18] 'Biotopes' – the smallest unit to be studied in the landscape – include hedges, roadside verges, drainage ditches, small brooks, bogs, marl pits, natural ponds, thickets, prehistoric barrows, and other small uncultivated areas. Inspired by the discovery of these often tiny biotopes, Dahl and his colleagues launched a research programme called 'Appearing and Disappearing Landscapes: The Dynamics of Small Cultures'.[19] Their aim was to find out what it could mean to design a city in which flows of water, plants, and animals were given equal status with those of people, goods, buildings, and information. What would it mean for infrastructure planning, they wanted to know, if cities and city regions began to reconceive themselves as elements of a bioregion?[20]

In parallel with this enquiry in Norway, then learned some enlightened cities, such as Toronto, have already started to put the interests of these natural assets ahead of traditional planning priorities such as transportation infrastructures. For Toronto, the practical way to reorder priorities is to put foodsheds and watersheds at the top of the agenda – a focus in design terms on 'reactivating the existing' – adapting and enhancing what is already there rather than continuing to accelerate capital and resource intensity. One of the Norwegian team's researchers, Alex Walls, calls this approach 'dirty' sustainability – giving priority to low-cost, hands-on solutions rather than high-tech ones.

I asked, a bit earlier, which country – Norway or India – has the most advanced and resilient infrastructure. As you may have guessed, my conclusion is that high-speed, high-entropy transportation systems take a country back to the past. They are not the way of the future. This is not to deny that there are many ways in which use of existing infrastructure, such as India's amazing train network – or, for that matter, the trip from Oslo to Stavanger – can be enhanced. But, looking elsewhere, I discovered that truly transformational changes in transport ecologies are already emerging. In India, showing the way to other cities, Chennai has embarked on a major programme to pedestrianize its roads; 60 per cent of the city's transport budget will be dedicated to non-motorized transport. Its municipal corporation, the oldest in India, is creating a network of footpaths, cycle tracks, and greenways to encourage residents to walk or cycle and to ease the passage of human-powered transport like cycle rickshaws and pushcarts. Critically, the new policy prohibits the construction of flyovers that could prevent parallel pedestrian infrastructure from meeting the right standards. It's a major shift in emphasis; big Indian cities typically allocate about 2 per cent of their budget to non-motorized transport.[21]

WILL IT MAKE ME SWEAT?

At a workshop in Delhi, during the UnBox Festival, I posed the following question to a group of twenty design, transport, and city development professionals: what new products, services, or ingredients are needed to help a cycle commerce ecosystem flourish in India's cities, towns, and villages?[22] The answer was: a lot – and it's not just about the bikes. We discussed the need for an online catalogue of products and business models to aid decision support. We learned that micro-finance for independent vendors should be a priority. Traffic architectures, hygiene regulations, and the disinterest of municipal authorities were an obstacle. Opposition from place-based retailers was also an issue. Topography and climate could not be ignored.

As the to-do list grew, the scale of the challenge seemed ever more daunting – but a strange thing has happened. The obstacles we identified in Delhi seem less daunting today than they did a short while ago. In China, 'battery-bikes' are outselling cars by four to one. Their sudden popularity has confounded planners who thought China was set to become the next automobile powerhouse. In Europe, too, e-bike sales are escalating. Sales have been growing by 50 per cent a year since 2008, with forecasts of at least three million sales in 2015.[23] Is this the start of a system-wide phase shift in transportation? I have the strong impression that a cloud of discrete but related developments is converging. In the background, a combination of energy costs and economic insecurity adds urgency to the need for change. At street level, myriad innovations in hardware, systems, and business models are giving us the component parts of the ecosystem we yearned for in Delhi. A profound transformation to the mobility profile of modern cities no longer feels like a dream.

CALORIE-COUNTING CITIES

A small project in Vienna confirmed my intuition that something big is afoot. I heard by chance about a piece of software that answers the question: will riding my bike from point A to point G make me sweat? A small firm called Komobile has developed a decision-support tool that will calculate the amount of nutritional energy the human body will need for a prospective trip by regular bike or e-bike.

Its project is part of a broader drive in Austria to promote e-bikes as 'range-extenders' to conventional bikes and increase two-wheeled commuting. The idea is to combine information about the mass of the vehicle – including its cargo and the mass of the rider – with data about the inclines, detours, and headwinds that increase the rider's body energy consumption. With this information to hand, riders can determine in advance whether the external energy provided by the electric motor on a pedelec is needed. The topographical data needed already exists in many digital maps. As one of Komobile's designers, Martin Niegl, explained it to me, *isohypses* are points of equal altitude that, when joined together, become the wavy black lines on maps we use when walking across country. Transport planners also use *isochrones* to denote lines of equal travel time. Komobile's innovation is *to add isoenergetes* – lines that plot units of equal energy consumption into a geographical information system (GIS) for metropolitan Vienna. The idea is to inform a rider how much effort will be needed to climb steep hills in the western part of the city, for example. For an unassisted cargo bike, this is crucial information. Komobile then discovered a missing piece of information – headwind speeds, which on a heavy bike can be just as taxing as inclines. In Vienna, Komobile located a source of real-time wind speeds in the city's meteorological office, only to learn that such measurement devices are located 10 m (33 ft) above the road surface. To be accurate, real-time wind speeds at surface level are needed. Adding a platform that can collect windspeed data in real-time from small devices around the city has been added to the to-do list.

MASS TRANSIT

Komobile's next task is to persuade city managers to embrace calorie-planning with the same confidence that they now plan time and space. Lightening up the movement of cargo around cities is the ideal place to start. Enormous amounts of energy are wasted shipping objects from place to place. An example from The Netherlands:[24] of the 1,900 vans and trucks that enter the city of Breda (pop: 320,000) each day, less than 10 per cent of the cargo being delivered really needs to be delivered in a van or truck, and 40 per cent of van-based deliveries involve just one package. An EU-funded project called CycleLogistics calculates that 50 per cent of all parcels delivered in EU cities could be delivered by cargo bike. Germany's Institute of Transport Research is even more ambitious: it reckons 85 per cent of all deliveries in a city like Berlin could be made by e-bike. Its finding was based on the experimental deployment of a so-called 'Bentobox' approach in which additional city hubs were able to coordinate distribution of goods. Cargo bikes need not be limited to lightweight packages. As the Belgian journalist Kris De Decker has discovered, fast two-wheeled cargo cycles have a load capacity of up to 180 kg (396 lbs); slower vehicles with three or four wheels can easily take 250 kg (550 lbs). Using a tandem configuration and/or electric power assistance can raise the load capacity even further, to about half a ton.

As Ivan Illich pointed out in *Energy and Equity* in 1973, the metabolic efficiency of a human on a bicycle is remarkably good. Measured in terms of calories expended by the traveller, the conventional bicycle is by far the most efficient means of human locomotion. To travel 1 km (½ mile) by bike requires approximately 5–15 watt-hours (w-h) of energy; the same distance requires 15–20 w-h by foot, 30–40 w-h by train, and over 400 w-h in a car with one occupant. According to ExtraEnergy's tests over several years, an average pedelec uses an average of 1 kilowatt-hour (kW-h) per 100 km (62 miles) in electricity. A car with an internal combustion engine uses fifty times more – at least 50 kW-h per 100 km (62 miles).

Translated into money, the difference is astonishing. It costs less than 1 US cent per 1.6 km (1 mile) to ride an electric bike or scooter.

A car, by comparison, costs *fifty times more* – 54 cents per 1.6 km (1 mile) according to the American Automobile Association once licence costs, insurance, registration, maintenance, and other costs are factored in. Kris De Decker reckons that, once all system costs are included, a cargo cycle can be up to 98 per cent cheaper per 1 km (½ mile) than four-wheeled motorized alternatives. Some e-bikers reckon that electric bikes can have a smaller environmental footprint even than pedal-only bicycles when the energy costs of the food needed to power the rider are added. If the rider eats a typical Western diet, about ten times more primary energy goes into the production of her food than is absorbed by the body when the food is eaten. Our metabolisms convert food energy into work with a conversion efficiency of about 25 per cent. The net result? For every unit of human energy used to pedal a bike, about forty times as much was expended upstream. Even including the energy needed to manufacture and recycle the batteries, e-bikes can end up consuming from two to ten times less fossil fuel energy than their human-powered equivalents.

THEY MAY BE CHEAP – BUT ARE THEY CLEAN?

Even if e-bikes *are* fifty times more efficient than a car per kilometre travelled, is the energy used to power an e-bike clean? Critics argue that if hundreds of millions of pedelecs were to be charged from the grid, the result will be more emissions because such a large proportion of the energy comes originally from dirty power stations. Policymakers in Austria are addressing this very real difficulty, too. They are deploying fiscal measures that incentivize pedelec users to use only renewable energy. In 2011, for example, a national subsidy for the purchase of company pedelecs was approved – so long as the company used only verifiably green electricity. What about the batteries? Probably 80 per cent of all e-bikes sold around the world use lead acid battery packs (they are called 'battery bikes' in China) – but these are indeed a suboptimal solution. Quite apart from the toxicity of battery production, the weight of lead needed to propel a bicycle for a decent 40–50 km (25–30-mile) range is a disincentive to regular use. From a sustainability perspective, lithium-ion cells are the better solution. Their high energy density means that only a lightweight pack is required. Trouble is, lithium batteries are expensive, and some sceptics warn of an imminent lithium shortage; lithium is the fossil fuel of the future, they say. The favoured solution to the battery conundrum in Europe is to change the business model. Hiring a battery should be cheaper than buying one, but at the moment extraction of lithium is still cheaper than recycling, so manufacturers have no incentive to invest in take-back. They make more money selling them as consumables. A ban on battery sales would reverse these perverse incentives. Anticipating this switch, the Swiss company Biketech supplies pedelecs without batteries; the rider hires them for around 15 euros per month. ExtraEnergy anticipates that this price will stabilize at about 12 euros a month in the medium term.

A shift to e-bikes is not just about reducing energy and cost. From a system perspective, they can also support huge numbers of livelihoods. Service providers and artisans were heavy users of cargo cycles during the first half of the twentieth century and now – as formal jobs disappear and transport costs rise – myriad small-scale traders are rediscovering neglected models, and reinventing them using today's new tools. As I have seen and marvelled in India over many years, bicycles enable a huge number of livelihoods in the formal as well as informal economy.[25] An e-bike can enable tradesmen, artisans, and service providers to start a business with a much lower investment, and to operate it at considerably lower cost.

One reason India is the world leader here is that the country has not yet modernized its bicycle commerce out of existence – but Europe is now chasing hard to catch up. The EU's GoPedelec project aims to help a vast army of traders migrate from their white vans to two-wheelers. In Germany, in an initiative called Ich ersetze ein Auto ('I replace a car'), forty iBullitt Pedelec Solar cargo bikes are being tested in nine major cities including Berlin and Düsseldorf. Pedelecs carrying loads up to 100 kg (220 lbs) and with a 90-km (55-mile) radius have running costs 85 per cent lower than a car or van. Germany's government takes the project seriously; its technology research is based at German Aerospace Center (DLR). In hilly

Stuttgart, the city has made pedelecs available for school caretakers and technicians who maintain public lighting around the city. In Austria, the city of Vorarlberg has made 500 pedelecs available to city workers at a reduced rate; in return, riders were asked to report about their riding behaviour on a regular basis.

Although cargo bikes are far less expensive than buying and running a van, they are still prohibitively expensive if you're starting up as a self-employed trader. In the Global South, where a lack of finance is not a new problem, a variety of financial innovations has emerged to meet this need – from bicycle micro-credit in Rwanda to the donation of recycled bikes to poor citizens in Chile. In Lagos, small traders, having founded their own credit system, now have their own banks.

Many citizens need to move packages from time to time, but not often enough to justify buying a cargo e-bike. To deal with that issue, the LastenRad Kollektiv, in Vienna, rents out cargo cycles to individuals who want to transport something big or heavy and prefer not to use a car. And in Israel, a platform called Tel-O-Porter makes goods trailers available as an addition to existing bike-share schemes. Tel-O-Porter's customers right now tend to use it for groceries and running errands, but the system is rugged and durable enough for work-related transportation, too.

DISTRIBUTED INNOVATION

My other reason for suspecting that all things e-bike are a mass movement, and not just a fad, is the scale, energy, and creativity of a worldwide network of innovators. The boundaries of technology, and new applications, are being pushed by many thousands of individuals rather than by well-funded research labs or big corporations. The Scottish-born CycleHack is active in fifty cities. The largest online forum, endless-sphere.com, boasts more than 13,000 topic threads and some 221,000 posts. This treasure trove of hands-on innovation will show you how to build e-bike packs from hacked-apart power tool batteries, or invite you to compete in making the lightest, most powerful, or cheapest e-bike. The scope of endless-sphere is constrained to some extent by language – it's all in English – but here, too, a solution is in the works: plans are afoot to build a pan-European blogging platform in multiple languages.

The e-bike movement also aspires to extend its embrace beyond the young and fit. Creative ways to engage car-bound doubters crop up everywhere. In Tanzania, free lessons in cycling skills are given to refugees. In Albania, bikeless citizens are serenaded by a movement called Shining Cycle Culture. In Ljubljana, whose citizens seem to dislike getting their hands dirty, a network of repair shops is being developed. Worried that your costly e-bike will be stolen? In Germany, a new ChargeLockCable system combines charging with theft protection at the same time. The system is being tested in three pilot regions: Tegernsee, Schliersee, and Achensee.

What about the proliferation of different technical solutions; will too much diversity limit the e-bike movement's ability to scale? There are seventy-three different charger plugs out there, for goodness sake. Fear not: the Germans are onto that, too. Design work on a standardized power connector – a USB plug for e-bikes – commenced in 2002 and electrical, data protocol, and mechanical definitions were published in 2011. What about space? Won't e-bikes compete with cars and pedestrians as their numbers multiply? The Dutch are showing the way on that one. They're building a national network of sixteen bicycle expressways at a cost of 80 million euros. The design specification includes no junctions or traffic lights; smooth asphalt; 4-m (13-ft) wide lanes to allow for easy overtaking; wind barriers on bridges; and even roofing on particularly exposed stretches. The new bike routes are integrated with public transport interchanges.

FROM BEDLAM TO GOVERNANCE

Every city's mobility ecosystem contains a multitude of economic actors: freight businesses, taxi drivers, courier companies, retailers, hotels and restaurants, emergency services, and street traders. How on earth is a city to persuade all these actors to head in the low-calorie direction signalled by e-bikes? In response to this challenge, a number of innovations in governance are emerging. The EU's

GoPedelec programme, for example, runs Municipal Decision Maker Workshops in Austria, Germany, Hungary, The Netherlands, Italy, and the Czech Republic. At each session local officials get to try out a pedelec, and learn about the hard and soft aspects of pedelec-friendly infrastructure. Best practice from other regions is shared at each event. In Denmark, a programme called Kickstand Policy Training helps diverse groups of local stakeholders work towards a shared vision for cycling. Kickstand's two-day courses bring together professionals from planning and design, local government, traffic engineering, urbanism and architecture, public health, tourism, economic development, neighbourhood and community groups, business associations, real estate interests, law enforcement, and school boards. The thinktank Embarq, together with UN-Habitat, has published an excellent five-step guide to setting up similar multi-stakeholder forums for urban mobility. Its 'Quick Guide: Establishing a Multi-Stakeholder Forum for Urban Mobility' tells interested cities how to improve institutional integration and ensure adequate representation of diverse actors. Another network, the non-bicycle-specific Cycloop, also provides the facilitation and action research needed to foster multi-actor collaboration.

As the writer Robert Neuwirth reminds us, diverse, fragmented economies are more resilient than hyper-connected global ones. Economic power in the developing world rests on millions of the small-scale businesses, family farms, local traditions, and extended social and regional networks that resilience experts are advocating as novelties here in the North. They may or may not be part of the legally recognized economic structure, says Neuwirth, but 'what happens among all the unregistered street markets and roadside kiosks of the world is not simply haphazard. It's a product of intelligence, resilience, self-organization, and group solidarity. It follows a number of well-worn though unwritten rules. It is, in that sense, a system.' This raises the question: can a networked mobility ecosystem be designed?

CLOUD COMMUTING

A two-year project in Belgium called Mobilotoop thinks it can. Inspired by the concept of 'cloud commuting', Mobilotoop proposes new relationships between people, goods, energy, equipment, spaces, and value. The seven signs on one small van describe the Mobilotoop approach: 'Taxi', 'Pick-up', 'Delivery', 'Assistance', 'Vendor', 'Security', 'Rental'. Seven functions are provided by one vehicle using multiple mobility 'assets' to support a wide variety of services. As imagined by Mobilotoop, the van, when coupled with a pay-per-use leasing framework, and radically distributed computing, becomes an element within an asset-light mobility ecosystem. In asking, 'How will we move in the city of the future?', Mobilotoop does not worry too much about the design of vehicles. 'Cloud commuting', in this context, is about accessing the means to move when they are needed (such as the micro-van) rather than owning a large heavy artefact (such as a Tesla) that will sit unused for 95 per cent of the time. Mobilotoop's design focuses on enhancing connections between people, vehicles, places, and services as a single ecosystem that generates new mobility solutions dynamically, and continuously. With a focus on connections that make us not just faster but also closer to one another, Mobilotoop is a system that enables us to become mindful in the ways we use time, space, and physical resources.

Mobilotoop envisions a mobility culture in which every ride is an encounter, every traveller an entrepreneur. Mobile media, flexible vehicle designs, and adaptive infrastructure enable everyone to be a user and a supplier of mobility services. Every commuter can deliver a package on her way to work. Every walker might collect sensor data about the quality of the pavement surface or the air. The electric motor on a pedelec might be used to drive a balcony hoist. In Mobilotoop's imagination, radically adaptive use is not only about cash transactions. A borrowed vehicle properly used and returned – or a service well executed – adds to your reputation as a sharer. This enhanced reputation gives you access to use-credits, discounts on services, or the use of other vehicles, equipment, and workplaces.

TOURISM, TOO

Sister Monika-Maria, one of forty nuns who live in a sixteenth-century convent in Wernberg, Austria, has guided us barefoot on a circular

'Path of Consciousness' over lush Carinthian meadows. Every few hundred metres, we stop for a short discussion about man's changing relationship with nature. Back in the convent's enormous garden, Sister Monika-Maria helps us gather armfuls of fresh herbs. We sprinkle these onto the cheese that comes from the nuns' cows, and is spread on the fresh-baked bread made from wheat they also grow themselves.

Wernberg is just one among a large number of locations in this part of Europe that can be visited along the 785-km (488-mile) hiking route called the Alpe-Adria Trail, which spans three countries and follows paths developed by walkers over hundreds of years. A Slovenian woman I meet organizes walking holidays called Apiroutes along part of the trail whose focus is bees and beekeeping. Slovenian beekeepers are known to be especially healthy, she tells me; they can live to a ripe old age and 'retain their physical energy and clarity of thought to the very end'. Reaching promptly for my diary, I am torn between an apitherapy study camp and a course in an api brewery on how to make honey wines.

So green tourism is alive and thriving, right? Well, yes and no. Examples like Bee Routes are a welcome alternative to a model of mass tourism that has seen coastlines concreted over from Spain to Mexico. Walking and camping are of course preferable to a tourist from a rich country stepping off a cruise liner for the day: she can use as much water in 24 hours as someone who lives there uses in 100 days. But green travel is a tiny speck in the bigger picture. Package tours account for 80 per cent of journeys to so-called developing countries, for example, and their business model is grotesquely exploitative. Destination regions receive 5 per cent or less of the amount paid by the traveller; 20 per cent on average remains in the country of origin, 37 per cent goes to the airline, and local intermediaries and investors take a substantial cut of what's left.[26] For local people on the ground, the injustice is absurd: if I were to pay US$1,200 for a week-long trek in Morocco's Atlas Mountains, just $50 would go to the cook and the mule driver who do the work.[27]

Travellers are not unaware of the problems with mainstream travel; 70 per cent of us would 'consider a green option' when planning a trip.[28] But there's no shared definition of what sustainable tourism actually means, and a bewildering variety of words and labels amplifies the confusion: hundreds of green-coloured websites talk about Sustainable Tourism, Responsible Tourism, Slow Travel, Nature Tourism, Green Tourism, and EcoTourism. They commit merely to 'minimize' negative impacts. There are no binding targets, and no governance of this vast and fragmented industry. The result is an empty promise to leave visited locations 'as unspoilt as possible'.

Twenty years ago, when the concept was first mooted, many people hoped that community-based tourism and ecotourism could be the basis of a new relationship between visitors and their hosts. The spectacular growth of sites like AirBnB prompted some optimism that peer-to-peer travel, enabled by the internet, might finally change mass travel for the better. There are three reasons these hopes have not been realized. The first is that tourism remains, at its heart, a form of consumption: we pay for an experience, not for a living relationship. Why else would the industry describe the communities we visit as 'destinations' or 'products'? Second: a commitment to 'do less harm' has no solid meaning in a business that keeps on growing; on the contrary, compound growth leads unavoidably to the occupation of untouched places. (The more untouched they are, the easier it is to package them as 'green' destinations.) A third dilemma is that tourism, however green, can never bring about environmental and social justice on its own. If biodiversity and social justice are to thrive, *all* the economic and social actors in a bioregion need to participate – with its long-term health as their shared goal. A consumer-oriented global industry can never be a means to that end. My own take is that while resource-sharing for travellers is cool, most new internet platforms empower the traveller – but not the destination. When P2P (peer to peer) travel start-ups pitch their idea to investors, the language is that of mass tourism: 'destination', 'product', 'targeting', 'the travel space', 'locals'. Moreover, the P2P model has to deliver large and growing numbers of transactions to satisfy investors; unique, respectful, and equitable relationships between hosts and visitors are dismissed as 'lifestyle' features – nice, but peripheral.

Some travel sites are making real efforts to move beyond the language of 'do less harm'. The impressive growth of platforms such as WWOOF (it stands for World Wide Opportunities on Organic Farms), HelpX, and the French Mission Culture & Communauté has tapped into a desire of many young people to contribute meaningfully to farming communities and the regeneration of living systems. These sites list organic farms, ranches, and lodges that invite volunteer helpers to stay with them and work in exchange for food and accommodation. In the typical arrangement, the helper works an average of four hours per day and receives free accommodation and meals for her efforts.[29] WWOOF, which is attracting 100,000 new members a year and represents 14,000 farms in more than 50 countries, is beginning to organize itself along bioregional lines: in Sweden, there's a WWOOF representative in each of the country's 25 regions. In small but significant ways, these projects exemplify a new narrative of living relationships, of connectedness, of respect for – and connecting with – The Other.

LEAVE-THINGS-BETTER TRAVEL

Although mobility grown into a vast global industry, it has done so at huge cost: the greedy use of space, matter, energy, and land just to move around. As the oil-based economy draws to a close, resource-intensive movement on this scale is no longer viable. The good news is that profligate mobility of this kind is not a universal need. Our natural inclination in pre-industrial history was to move only as much as we had to. Even today, 40 per cent of the time we spend travelling, across all cultures, is spent walking or waiting, and for the lion's share of longer day-to-day distances, as I've described here, an elaborate ecosystem based on bicycles, some of them power-assisted, will meet most of our need to connect and transact with each other using 5 per cent or less of car- and train-based systems. As for the longer trips we will still want to make, we'll simply have to spend more time and move more slowly to do so. Movement will be more like a place, than a means to an end, and there will be plenty of ways to enhance it.

8 CARING:
FROM CURE TO CARE,
FROM ME TO WE

I was emboldened, upon arriving at the Mayo Clinic's Centre for Innovation, to learn from the conference host that 'people with deep domain knowledge do not make the best innovators'. I concluded that I was therefore well qualified to warn one of the top academic medical centres in the world, each of whose 60,000 staff knows more about medicine than I do, about the risk of catabolic collapse in the US health system – and what to do about it. My core proposition at the Mayo event was that peak oil, and peak fat, are transforming the logic that currently shapes the global biomedical system. Firstly, because the energy transition that's upon us will render one of the world's most energy-intensive systems unsustainable. And second, because until the medical system addresses the causes of illness with the same brilliance with which it addresses the effects, the population will continue to get sicker.

The main Mayo Clinic building is a vast silver facility that shouts two things: authority and energy intensity. If one Googles 'health' and 'energy efficiency', most results are about hospital buildings and attempts to render them 'greener'. But hospital buildings are just one element within a vast distributed system that is both materially heavy and energy-guzzlingly complex. At a practical level, most of the consumables within any hospital are oil-based – from analgesics and antihistamines, through heart valves, implants, and prosthetics, to ambulances and helicopters. But energy that you can measure, such as that used by buildings and suppositories, is only one part of the picture; the total energy demand of any business operation, including health ones, is four or five times more than is ever measured. A recent UK study, for example, found that 5 per cent of all vehicle movements on British roads are health-related. This energy blindness is significant; because the true costs of so many activities is neither perceived nor counted, no thought is given to their possible replacement.

My sombre words at the Mayo conference were met by a sea of blank stares. I was not offended: the medical world is preoccupied by other issues than the consequences of energy transition. The danger facing complex organizations such as the Mayo Clinic, nonetheless, is that, by postponing consideration of energy issues, it risks 'catabolic collapse' down the line. This is the situation, as described by John Michael Greer, in which, by the time a system realizes that its energy regime is not sustainable, the money, energy, and resources to do anything about it are no longer available.

PEAK FAT

From catabolic collapse I moved on in my talk to peak fat. I was perplexed at this fascinating conference by a weird imbalance. I saw several case studies about innovative ways to deal with consequences of the diabetes pandemic; by 2030, it's forecast that 438 million people will have diabetes worldwide – a 54 per cent increase on today's total.[1] The response of designers and doctors present was an array of Personal Health Planning tools, 'high-end wellness' services, superfoods, remote diagnostics, and more. But I heard almost nothing about tackling the *causes* of this grim disease. That oversight is system-wide in medicine. The Centers for Disease Control (CDC), for example, in a briefing about obesity, explains that these killer conditions 'result from an energy imbalance. Behaviour and environment play a large role…these are the greatest areas for prevention and treatment actions'. This is a bizarre statement. 'The environment' makes you fat? I was under the impression that *fat* makes you fat – and the correlation between the growth of fats in the food system and the growth of obesity and diabetes in the population, is not hard to spot There's even a word for it: the 'obesogenic environment'. Neither is it a secret that the producers and distributors of this killer fat are the junk food and soft drinks industries.

The medical system – from the CDC to the Transform conference – is focused overwhelmingly on downstream phenomena. Smart and expert physicians are working tirelessly to improve treatment protocols; huge efforts are also being made to improve safety in hospitals, and raise the quality of care received there. But among all these projects, references to transformation of the food system that is making hundreds of millions of people sick are rare to invisible. A speaker from a soft drink company added to the cognitive dissonance. This product developer told the conference that she is 'on a journey… to redefine what nutrition means' and that her employer is determined to 'double its food-end healthier portfolio'. I wanted to ask, 'Healthier than what?' The only way for this firm to make money in the processed food industry is by innovating as much as possible – but healthy food, fresh-from-the-field food, whole-until-the-kitchen food, loses quality with each 'value-adding' innovation process it is subjected to. In the absence of an announcement that it will close down its sugared beverage business, the company's food-end healthier portfolio can't possibly make the nation healthier.

A more plausible proposal from a global business was made by IBM's Dr Paul Grundy. The conference responded enthusiastically to his Patient Centered Medical Home. Grundy is one of the world's leading experts on preventive medicine, so he did not peddle platitudes. His model is about the integrated use of phone, e-mail and internet portals to enable one-on-one communication and collaboration between doctors and patients. Each occupant of the PCMH will enjoy an ongoing relationship with a doctor for continuous and comprehensive care and – the killer line – 'we know that somebody who has a personal physician will cost us about a third less'. The IBM scenario is enticing – but for me it suffers from two flaws. The first is cost. 'A third less' will be welcome to those fortunate people who are in work, and are therefore insured. It will be unaffordable for the majority of the population who need it most. Of course if IBM's share of the pie, and that of highly paid doctors, could be removed from the equation, the story would probably be different – but as it stands, IBM's smart home is for the top 5 per cent, not the bottom 95 per cent. But the even bigger drawback, for me, is that the design concept is about one-to-one connectedness through technology – not the social, embodied, eye-to-eye kind of connectedness that sustains healthy communities elsewhere in the world. I'll come to that a bit later.

MEDICAL INDUSTRIAL COMPLEX

Blithely indifferent to considerations of cause, investor interest in downstream medicine is, if anything, growing. Their latest wheeze is to combine the allure of biomedical research with gigantic real-estate projects. So-called biomedical clusters, which concentrate research institutions, universities, hospitals, biotechnology spin-off companies, and pharmaceutical firms in the same geographical area, have been growing like topsy. Their number includes Biolake in Wuhan, Genome Valley in Hyderabad, Health Valley in Geneva, BioTOP in Berlin, Genopole in Evry, Brainport in Eindhoven, and many more. Sweden's entry, Biomedicum, is one of the largest building sites in Europe at the time of writing; its 55,000 squares will house 1,700 researchers.[2] Europe even has its own research hub, Health ClusterNet, whose mission is to 'increase general understanding of the "health is wealth" relationship'.[3]

A more honest wording would be 'the *promise* of health and wealth relationship'. These huge investments are based on a scientific impossibility: that technology will help us cheat death. Investors are untroubled by this quibble. So beautiful is their business model that they don't have to *deliver* immortality – they just attach the *promise* of future health breakthroughs to massive real-estate plays – and sell on their stake a few years down the line. Participating universities are also exempt from being judged by results. Few of the researchers in bioclusters are paid salaries by their university – most of them hustle for their own funding from third-party sources – but for every researcher sitting in a lab, the university receives on average a US$100,000 overhead paid by whoever is sponsoring the research project in question.

Bioclusters have been great business for construction firms and investors, but a day of reckoning is imminent. The bubble will burst when the poor returns on all this investment are revealed. According to the Rockefeller Foundation, 'game-changing advances in science' represent just 10 per cent of the key trends impacting health futures.[4] In Canada, the primary factors that shape health have been found not to be medical treatments at all; when researchers listed the actions that would make the biggest difference to the health of the country's citizens, not a single medical treatment, drug, or lifestyle choice appeared in the top fourteen. Having combed through decades of research and hundreds of studies, they found that the primary factors shaping the health of Canadians were their lifetime living conditions: early childhood, education, current employment, income, housing, community cohesion, and so on.[5] In the USA, it's even worse: receiving medical treatment is the third leading cause of death[6] – and receiving the bill must come close behind. As a wag at the Mayo conference put it, 'medical science has made such tremendous progress that there is hardly a healthy human left'.

Even as a business, the Medical Industrial Complex is a poor performer. The anthropologist Joseph Tainter reckons that the 'productivity' of the United States health-care system – its capacity to improve life expectancy – has been declining since the 1930s. This is because inexpensive diseases and ailments were conquered first – the basic research that led to penicillin cost $20,000 – and each condition thereafter has proved more difficult and costly to resolve.[7] The writer Ramez Naam even proclaimed a law – Eroom's Law – to describe this productivity slowdown. Unlike Moore's Law in computing, by which performance improves continuously, Naam discovered that the rate of new drugs developed, per dollar spent by the pharmaceutical industry, has dropped by a factor of 100 over the last 60 years; this is why Naam's 'law' is Moore's Law spelled backwards.[8] As its efficacy has declined, the overhead costs of the Medical Industrial Complex have skyrocketed: in the US alone, $400 billion is spent annually on an immense army of medical administrators, laboratory staff, specialists, insurance agents, government officials, marketing and advertising creatives, lobbyists, bonuses for CEOs, and profits for shareholders. Even in the US, the world's richest country, negative metrics like these cannot be ignored for ever. Funding for biomedical innovation at more than 2,500 universities across the nation has surely peaked.[9]

FIVE PER CENT HEALTH

My argument throughout this book is that we need to grow social support systems – including health-care ones – that can flourish using 5 per cent of the resource costs per person that we have now.
I know this sounds like a fantasy, but consider this: in Cuba, where food, petrol, and oil have been scarce for sixty years as a consequence of economic blockades, its citizens achieve the same level of health for only 5 per cent of the health-care expenditure of Americans.[10] Bangladesh, one of the world's poorest countries, is also well known for impressive improvements in a range of indicators, particularly child mortality. These achievements have taken place despite relatively low levels of spending on health, but with substantial innovations in community-based service delivery, health extension worker programmes, traditional birth attendants, and programmes to improve treatment of diarrhoea.[11]

Across the Global South, although billions of people are geographically and economically excluded from direct contact with medical professionals, a large number of community health workers display the same depth of knowledge as clinic-based experts in the US or Europe. Ninety per cent of primary health care is provided in the community with the use of low-cost medicines and equipment.[12] In northern countries, too, 5 per cent health care already exists where workers with less training than doctors operate in communities, and away from big hospitals. So-called nurse practitioners, or physician assistants, can perform about 85 per cent of the work of a qualified family doctor. And we, the patients, are perfectly happy with the service received: when different nurse practitioner schemes around the world were reviewed by the *British Medical Journal*, patients treated by nurses were found to be more satisfied, and no less healthy, than those treated by doctors.[13] I heard about an especially striking example at the Mayo conference. In New Mexico, community health workers are paid $10 an hour to work on the front line of a highly effective campaign against common diseases such as hepatitis, asthma, and substance misuse. In some of the state's prisons, inmates, after a ten-week training course, are also proving highly effective health educators to their peers – and they are paid nothing. 'Doctors', said one professional at the Mayo Clinic, 'are grossly overemphasized.'

FLIPPING THE PYRAMID

For a long time, my stories at health industry events about Cuba, or convicted felons as exemplars of a more resilient health system, proved to be a hard sell. But the tide is turning in a big way. At a recent care conference in Eindhoven, for example, a Cuban-style strategy was advocated by a mainstream industry leader with real financial clout. Roger van Boxtel, CEO of a big Dutch insurance company, Menzis, which has two million insured clients, used the image of a pyramid to illustrate his company's plans to redirect spending. The top of the pyramid, where doctors and costly hospitals treat acute patients, currently absorbs the lion's share of health spending. Van Boxtel then turned the pyramid upside down: 'From here on, we will focus our resources upstream,' he told a shocked room, 'on prevention, and on helping people manage their own long-term conditions.' When I asked the head of a huge hospital, on the same panel, what he made of this startling redirection of resources, his rueful reply was that 'if he says so, that's the way things will go'.

Menzis does not propose to do away with hospitals altogether. But it does intend to reduce costs radically by focusing common procedures at a small number of preferred suppliers. It will send all patients for hip replacement, for example, to a single clinic that already performs 700 hip operations a year. Logically, it is hard to see why Menzis's inversion of the Follow-The-Money principle should stop at Dutch borders. An international patient visiting India can save 70–80 per cent on the average cost of a similar procedure back home. Hip replacement surgery at the top-rated hospital in India costs US$5,000.[14]

The consequence of Roger van Boxtel's cost-reducing strategy is a radically different patient experience. Health and wellbeing will no longer be something 'delivered', like a pizza, by distant suppliers. On the contrary: the focus is shifting to the nurture of mutually supportive relationships between people in a real-world context, away from big medical institutions. This is no small shift of emphasis; the 'delivery' metaphor is pervasive in the developed world's systems. Cuban-style 5 per cent health is not about a U-turn back to a pre-scientific age; it's about focusing resources and creativity on the 95 per cent of care that happens outside the medical system already, today. It's about reimagining the 'health space' as a social and ecological context which, like a garden, needs to be cared for – collaboratively.

PEER-TO-PEER HEALTH

'We cannot be healthy alone.' Graham Leicester, director of the International Futures Forum in Scotland,[15] points to the considerable evidence amassed by health psychologists that a sense of social support is the best buffer against illness; strong social networks decrease the length of recoveries and reduce the probability of mortality from serious diseases.[16] People with higher levels of support recover faster from kidney disease, childhood leukaemia, and strokes, have better diabetes control, experience less pain from arthritis, and live longer.

A number of new internet platforms have emerged to augment this social dimension of health. One approach is the mass polling of patients to determine what actually works. An organization in the United States called Lybba, for example, has developed a Collaborative Chronic Care network (C3N) that combines social and idea-networking functions, 'remote wellness tools', and medical content.[17] Another website, CureTogether, is able to ask six thousand patients what treatments work best for each of them.[18] The aggregated results are fascinating: totally free remedies, such as 'exercise' or 'masturbation', are plotted on the same effectiveness versus cost chart as with dozens of drug therapies. On an even larger scale, a movement of 'self-trackers' called The Quantified Self (QS) are using smartphone apps and assorted custom-built devices to monitor patterns of food intake, sleep, fatigue, mood, and heart rate.[19] Although the QS value proposition of 'self-knowledge through numbers' plays fast and loose with the world's spiritual and wisdom traditions, the QS movement has grown quickly on the back of a simple promise: the combination of individual data and large group datasets will help people make their own better choices about their health and behaviour.

DOCTORS AND DEMENTIA

An overemphasis on doctors may not make us healthier, but it's a hard habit to break. The cultural and economic lock-in of mainstream medicine was brought home me by the G8 Dementia Summit in the UK, in 2014. It so happened that, during the Dementia Summit, I was spending time at a care home in England that looks after people with dementia and terminal illness, and their families – including, this time, mine. Teams comprising three carers and a qualified nurse worked twelve-hour shifts. Their hour-by-hour duties included helping people eat and drink; changing clothes and bed linen; helping people shower and clean; cutting finger- and toenails; helping people use toilets, bed-pans, sanitation pads, and commodes; filling in forms; and attending staff training sessions. A lot of the time – and I mean a *lot* – carers would sit and talk with the residents, reassure them, read books or magazines together, or simply hold their hands, or hug them. Twice a day, it's true, one of the nurses would tour each wing to give medicines to the residents, and a few residents needed more intensive medical attention. But I gained the strong impression that 95 per cent of this demanding, time-consuming, and emotionally draining work involved caring – not doctoring, and not 'curing'. In my family's case, we saw a doctor just twice during those weeks; on both occasions, they were there to authorize a medication that the nurse practitioner was not allowed to prescribe on her own. These brief appearances by doctors were rare interludes in the perpetual presence of quiet, attentive staff in the home.

During a brief respite, when I turned on the television for the first time in weeks, it was to see the British prime minister addressing a room full of people clad in smart suits and name badges. Speaking in forceful, Churchillian style, the prime minister declared that 'we must fight dementia' and announced a £100 million global research project to find a cure. 'I know some people will say that it's not possible,' said Mr Cameron, 'but I will not be defeatist. With a big global push we can beat this.' Media coverage of the summit echoed the politician's rhetoric. The airwaves resonated with words such as 'cruel disease', 'time bomb', 'robs you of your mind', or 'horrific'. People suffering from dementia were labelled 'victims'. One BBC reporter, who delivered his piece to camera while standing beside a million-euro brain scanner, looked and sounded like a salesman for Siemens. Cutaway shots featured glowing, computer-enhanced images of peoples' brains – a visual language that amplified the myth that dementia is like a tumour that, given fancy enough technology, can be cut out.

As I watched the politicians do their thing, and the media theirs, I finally understood what many dementia sufferers and their carers had told me in the past, but which I had not really understood. The net effect of war-speak is to diminish social solidarity. Words like 'war' and 'fight' worsen the stigma attached to the condition, increase fear in the broader population, and leave millions of sufferers and carers more isolated than ever. Depicting people as helpless victims – as refugees from a lost battle – disrespects and undervalues the often admirable lives led by an infinitely diverse group of people. And all this for what? The likely outcome of this so-called 'race to identify a cure' for dementia will be the creation of a Dementia Industrial Complex. It will be run by and for the glossily clad experts in the prime minister's audience – and it will do little to support the low-paid care workers who support my family and thousands of others like it seven days a week. Neither patients nor carers were even represented on the much trumpeted World Dementia Council that was set up after the G8 Summit. Just fourteen months later, the following headline appeared in a British newspaper: 'Drug firms despair of finding cure and withdraw funding after a catalogue of failures.'[20]

Care is principally a time issue, rather than a throwing-money-at-research-and-technology issue. The better way for nations to spend money on dementia is therefore in the ratio: 95 per cent for Care, 5 per cent for Big Research. I do not pluck those numbers out of thin air. They correspond to the structure of the care economy we have now. More than a quarter of the US adult population have provided care for a chronically ill, disabled, or aged person during the past year for example, and that's more than 50 million people – a number that will grow as the population ages. In Wales, 340,745 unpaid carers – 11 per cent of the country's population – provide 288 million hours of care per year; local authorities provide 12 million hours.[21]

What would a shift of focus to support for care look like? In a project called Alzheimer100: Improving the Journey through Dementia,[22] I learned that this is not a complicated question. The one sure way to improve life for people living with dementia is to ask them directly: what practical steps might improve your lives? So we did. We worked over a two-year period with a wide variety of citizens: people with dementia, carers, support and voluntary groups, researchers, doctors, and nurses. We used a variety of approaches to determine what the needed practical actions might be at different stages of what we came to call 'patient journeys'. From this process emerged a wide variety of anecdotal insights – but also, crucially, a shared understanding of what, for people with dementia and their carers, were priorities. The two most pressing priorities for them were first, the awful experiences people have when they first discover they have a problem; and second, end-of-life experiences. These two issues were both judged to be more than twice as important as assistive technology (AT). This is because the majority of AT applications are developed by university-based researchers for a category called 'the elderly' whose members they seldom encounter and whose true needs they cannot understand. A lot of R&D money has gone into 'proactive systems' that enable adult children to assess the health and wellbeing of their ageing parents while they – the children – are still at work; a Japanese-led research team has developed a robot to 'help care for the country's growing number of elderly'. The thinking behind these ideas is depraved, but tech companies are mesmerized by the vast market among working adults who would rather pay for automated care than provide it in person.

With that to-do list as its guide – first discovery of a problem, and end-of-life experiences – our team of researchers, designers, carers, and dementia patients was able to identify a shortlist of ideas for support services. In our case, these included ideas for a 'care concierge', a buddy system, and an eBay for time. In the event, we focused on the development of a prototype dementia signposting service to enhance connectivity among existing support ecology; this service has since been adopted and developed nationally by the UK Alzheimer's Society. The simple action of asking the people involved what would make a difference proved liberating on a wider scale. Following Alzheimer100, the UK's Design Council launched a Dementia Challenge whose stated objective was to 'get new solutions up and running and into the hands of the people who need them'. The sixteen service ideas shortlisted were relatively low-cost, low-tech, and people-focused solutions: online tools to connect relatives, friends, and professionals; a web-based service to help carers find part-time work; a ride-share service to facilitate daily journeys for people with dementia; a volunteer network to locate and safely return people with dementia who wander; a peer-to-peer platform for carers to help each other. The Design Council's shortlist did include the odd high-tech gizmo: a wristband to monitor the location of people with dementia included 3D accelerometers and RFID. But the tech component overall was modest.[23]

Between now and the year 2030, the number of people in Europe aged over seventy-five will double. At least half of these elderly people will live alone, and a growing proportion will suffer from at least one chronic disease that requires ongoing medical care. At the same time, state spending on public services in many countries is likely to shrink by 20–40 per cent in the coming years.[24] The conclusion is inescapable: traditional patterns of health service and elder care will not be sustainable in this future context. How, then, will our governments care for older people? The honest answer? They won't. They can't – or at least, not if the words 'care for' are interpreted to mean the complex, high-cost, energy-intensive, and institutionally focused kinds of medical care that absorb the majority of resources now. We need to reframe the words 'care for' and explore, instead, how best to build on the trust between people, built up by co-presence over time, that is judged by dementia patients and their carers to be more important than the delivery of services by third-party vendors.

A cooperative care ecosystem does not have to be invented from scratch; examples already exist. Fifty per cent of elder care provision in Quebec is cooperative, for example. And in Bologna, Italy, a study by John Restakis found that over 87 per cent of the city's social services are delivered via service contracts between the municipality and social co-ops. As a result, the city has experienced a dramatic growth in the number and variety of health and social services available, an improvement in the quality of care offered, and a lowering of the cost of providing these services.[25] In Italy as a whole, more than 14,000 social cooperatives – with a workforce of over 400,000, and an annual turnover of more than 9 billion euros – deliver a wide diversity of care services to over five million people. This co-op system is self-programmed to grow: the creation of one new co-op brings with it the obligation to set up another. In a practice known as the 'strawberry patch' principle, each new 'plant' is obliged to put out a runner and propagate at least one offspring. Inspired by this success, similar care co-ops have recently been launched in several other European countries.[26]

GRANNIES DUMPED ON THE MOUNTAIN

In China, where 430 million people will be aged over sixty by 2050, a novel form of elder care combines time banking with a volunteer programme. The idea, as described on Slate by Benjamin Shobert, is simple enough: adult workers receive a basic level of training on senior care from a government agency and then volunteer to provide elderly people with companionship and basic care services such as cleaning, shopping, cooking, counselling, and personal hygiene. When the volunteer reaches retirement age himself, the total number of hours he has volunteered so far become credits for his own care needs.[27]

In Japan, as I learned from David Pilling, nursing homes have a bad image; elders sent away are referred to as 'grannies dumped on the mountain' – a reference to an alleged practice in ancient times. Today, all citizens over forty are obliged to contribute to elder care insurance, and Japan has the highest provision of day centres for elderly people in the world. The system embodies the spirit of *ikigai*, a Japanese word that translates as 'a reason to live', something to keep mind and body active. 'We want to make sure old people have moments of joy,' Pilling is told, 'that they can eat great food and spend more time with the friends and family they love. We are less concerned with extending life than in maintaining its quality.'[28]

In the UK, in a membership model called Circle, local neighbourhood helpers provide on-demand assistance for elderly people with daily practical tasks. The helpers, who are technically supported by professional social workers, accumulate credits within a time-bank. For the older people, the benefits are not just practical ones. Participating in a Circle helps them to stay socially connected around shared interests and values – and not just with other older people; it also enables many to contribute to their local community. Each Circle is set up as its own Community Interest Company and run as a social enterprise; these receive investment from local organizations, principally local councils and housing associations.[29] A network effect is also evident in a platform called Dementia Friends, also in the UK; more than one million people have signed up to 'learn a bit more about dementia and the small things you can do to help people with the condition'. And in Greece, where a large part of the public health system collapsed following the 2008 crisis, many doctors, nurses, and pharmacists give their services voluntarily and for free in 'social health clinics' that provide free medical services, drugs, and vaccines for people without access to public health facilities.

At the time of writing, forty-two such clinics were active in different parts of the country.[30] Another network-enabled approach, this one in Canada, is Tyze Personal Networks. The Tyze platform, which is already being used by ten thousand people, helps connect the variety of people who cluster around someone in need of care: family, friends and neighbours, and professional care and health-care staff. Because Tyze connects carers to each other, and improves communication among them, it also reduces stress for friends and family members who often feel isolated and overwhelmed.[31]

The social care economy already exists – and not just in families. It also exists among the myriad support networks – formal and informal – that most of us only find out about when we need them. Much of the work we do in this social economy is unpaid, and it does not get measured as GDP – but it's nourished by other kinds of value than money: the trust, time, attention, wisdom, experience, and skills that we all contribute in caring for each other. Building on what is already there, this social ecosystem can be supported and improved in many different – and affordable – ways. New communication tools can improve coordination between professionals and the rest of us 'actors'. Different kinds of knowledge and skills can be connected using the power of networks. The ways we share resources – from diapers to buildings – can be enhanced by new services.

Just like food, it emerges, health is local and place-based, too. As the consequences of that sink in, an alternative to today's disease-centric biomedical system is gaining traction. In this new ecological model of health, the focus is shifting to the places and ecosystems we inhabit. This ecological model is no longer fringe. In the US, fifty-seven academic health centres – including Mayo, Allina, and Harvard – have joined an Integrative Medicine consortium.[32] And a clinician-based Academy of Integrative Health and Medicine (AIHM), launched in 2014, is based on the idea that health is determined primarily by the vitality (or otherwise) of food, water, air, and other ecosystems – not by procedures carried out in hospitals. The connection with food goes further: the AIHM is promoting the concept of a Health Commons in which the health of ecosystems, and the people who live in them, are addressed as one.

9 COMMONING: FROM SOCIAL MONEY, TO THE ART OF HOSTING

In a sleepy hamlet an hour from Bangalore, India, I encounter a group of villagers standing around a wide patch of *ragi* (a grain that is used to make dark bread), spread thinly over the road in a neat circle 6 m (20 ft) wide. Six chickens appear to be eating up the grain while the villagers watch and chat. Why, I ask, don't the villagers feed the grain in a trough? They laugh, good-humouredly, and then explain that the chickens are eating tiny maggots, smaller than our eyes can see, which need to be removed from the grain before it can be stored. It's a smart, low-tech solution to a practical issue faced by farmers everywhere. Back home, a Google search for 'clean bugs from grain' throws up the 'Opico Model 595 Quiet Fan Batch Dryer With Sky-Vac Grain Cleaner'. When it comes to doing more with less, I concluded, my village in Bangalore beats the Bay Area hands down.

The word 'development' is often used to imply that we advanced people in the North must help backward people in the South catch up with our own situation. But consider this: the average US citizen emits as much CO_2 in one day as someone in China does over a week, or a Tanzanian in seven months. Or this: a tourist from a rich country uses as much water in 24 hours as a villager who lives there uses in 100 days. And we want these frugal peoples to catch up with us? The word 'development' only seems to make sense to the people who are doing it to someone else.[1] From colonial times until now, rich people in the North have assumed without question that our way of life is more advanced than everyone else's. In particular, we have tended to view the people and ways of life that are already there as impediments to progress and modernization. This mindset has led to a tsunami of negative impacts as the global economy has grown. Today, ten million people a year suffer forced displacement from their homes and livelihoods to make way for dams, transportation systems, and waterfront developments. An international army of banks, government agencies, and property developers measure progress only in terms of private home ownership and sharply increased consumption. And, as itemized by the writer Charles Eisenstein and many others, this industry assumes without question that urbanization and transport intensity are signs of progress; that walking should give way to cars; that tower blocks are better than multi-family compounds; that concrete is superior to local materials; that paid-for insurance is better than mutual aid; and that fast-food restaurants are an improvement on subsistence agriculture. The result of all this development, in the chilling words of Maggie Black, is that 'millions of people are expelled to the margins of fruitful existence in the name of someone else's progress'.[2]

How times change. By a twist of energy-fuelled fate, the difference between 'them' and 'us' is now dissolving. The British writer Guy Standing coined a new word, *precariat*, to describe the growing proportion of citizens in rich northern countries for whom insecurity and relative poverty is the new normal.[3] In contrast to many of our parents, who enjoyed secure careers, health benefits, and pensions, the new normal for members of the precariat is agency work, zero-hours contracts, and uncertainty. I'm not talking about a marginal underclass here. The OECD calculates that half the world's workers – almost 1.8 billion people – already subsist in the precarious economy; by 2020, they project, two-thirds of all adults in the world will be economically informal. Even that startling percentage may be an underestimate. If one adds in all the people who do have 'proper jobs', but worry that they may not last, then the true membership of the precariat will soon be three-quarters of all working-age adults in the world.[4]

'What is it like to live in such a stripped-down, pitiless reality?'[5] For the rich-world journalist who wrote those words, the prospect of joining the precariat was clearly terrifying. But as a long-standing member of the world's largest and most unpopular club, I've come to a surprising conclusion: it's not all bad. The biggest positive is that the end of growth, in the real-world economy at least, seems finally to have arrived. Why would I welcome such a thing? Let me explain. Many technical experts argue, when pressed, that for our world to be sustainable it needs to endure a 'factor 20 reduction' in its energy and resource metabolism – to 5 per cent of present levels.[6] For much of my adult life Factor 20, as environmental researchers call it, has struck me as being beyond reach; its sheer implausibility turned me, for quite a while, into a confirmed doomer. But then I began to spend more time with people at the bottom of the pyramid for whom 5 per cent energy is a lived reality today. This is not to underestimate the many challenges faced by poor people on a daily basis – and it would be insulting of this writer to lecture poor people about their good fortune – but when it comes to living on modest resources, the poor people of the world are further along the learning curve than the rest of us. Many people I have met enjoy a decent and tolerably comfortable life on a tiny fraction of the energy and resource base that we are used to in the North.

The owner of a small guesthouse in Kerala, in southern India, opened my eyes even further. During a discussion about our parents, his jaw dropped when I complained about the cost of care homes in England. He literally could not believe that we, the children, would even consider putting our parents in a care home rather than in our own homes. Although the guy was dirt poor by rich-world standards, he took the supportive relationships that surrounded him totally for granted; he was looking after his mother in his home at the time, and fully expected his own children to do the same for him. Konpè Filo, one of Haiti's most popular journalists, reassured me that I was not fantasizing in a memorable interview. When he was asked, after the disastrous earthquake, whether Haiti was actually poor, Filo replied, 'It depends on how you define poverty and wealth – and who does the defining. I would actually say that Haiti is a rich country. We have solidarity and community. We're raised in compounds with common courtyards, and we know that what you have, you have to share with your neighbours. You stand in front of your neighbour's house and you ask, "Did you drink coffee already today?" You know that your success and your family's success depend on the community's wellbeing. That's the model we have.'[7]

I've also learned that a huge amount of healthy agriculture still exists. Eighty per cent of all farms in the world – 445 million of them – occupy 2 hectares (5 acres) or less[8] and in most cases these small-scale operations are living examples of the ecological agriculture that needs to take over from the 'production' kind. Yes, many of the farmers I've met face real and growing threats to their livelihoods – but 40 per cent of global food production still comes from diverse smallholder agricultural systems in so-called multifunctional landscapes, and an estimated 1.6 billion people still use woodlands as sources of livelihoods and income. Even as the bulldozers of production agriculture roar nearby, huge numbers of people still obtain building materials, fruits, nuts, mushrooms, honey, and medicinal plants from the forests that remain.[9] It's the same with urban agriculture. While a rooftop plantation in London or New York today is newsworthy, 800 million people in the South have been growing food in cities for decades; in sub-Saharan cities, 40 per cent of households are also urban farmers.[10] Whether it's inside cities or outside, this bottom-of-the-pyramid agriculture is more sustainable than our own: their ratio of energy inputs to the food ingested is about even compared to our 'production agriculture', where, as we saw in Chapter 5, the ratio is more like 12:1.

The same expert capacity to do more with less is evident in housing: poor people use far fewer resources to shelter their families than we do in the McMansionized North. As I described in Chapter 4, people living in favelas have ten, twenty, thirty times fewer square metres per person to live in than rich people do – but the social solidarity available in the rest of their neighbourhoods can compensate for a lack of private space. In material terms, poor people's construction is especially efficient compared to our own. In sub-Saharan Africa, for example, an ancient architectural technique called the Nubian Vault uses raw materials that are cheap, locally available, and ecologically sound. The major cost of the buildings is labour – but because the construction procedure is easily learned, and much of the work is done by the future occupants of the house, costs are kept low and money remains in the local economy. In contrast to our out-of-control Real Estate Industrial Complex, Nubian Vault construction is becoming autonomous and self-sustaining.[11]

Health care is another necessity of life in which the South is ahead of us – by example, if not by choice. Because poor countries cannot afford the doctor-focused, pay-per-procedure, treat-the-symptoms-not-the-causes medical systems that are on course to bankrupt rich countries, they focus – because they have to – on community-based health and prevention. Physicians are based in neighbourhoods, not in clinics or hospitals. Community health care is carried out by trained local people, not only by doctors. A lot of this social medicine is self-replicating, too; in countries such as Venezuela, doctor-teachers recruit and train health workers from among peasants and workers. The goal is to empower local people to provide 90 per cent of their own health-care needs. This is 5 per cent health in practice.

Poor communities are ahead of the game, too, with off-grid energy. As energy precarity bites in the North, we have much to learn from the ways poor communities procure, deploy, and share affordable, general-purpose, low-tech tools and equipment. The open-source technology movement is accelerating the development of distributed peer-to-peer energy production and local area networks; these permit energy flows from many to many, and are based on the voluntary participation of independent producers, households, or communities. There is huge scope for North and South to join forces. Local energy, as described by Kevin Carson, is 'the world's biggest coordinated DIY effort'.[12]

The South has much smarter cities than we do, too. They are not filled with Big Data and ICT consultants, it's true, but the streets of poor cities are sites of intense social and business creativity. Every time I go to India, for example, I am amazed by the prowess of the pavement-based engineers and fixers who look after the gadgets and equipment of a gadget-filled city: engines, television tubes, compressors, and other devices. In Delhi, near an office I shared for a while, hundreds of tiny workshops, plus sole traders sitting on the street, would sell and fix the hardware peripherals needed to keep our office running. Everything from toner cartridges to USB sticks was available; and in rows of gloomy but bustling basements, an amazing array of ancient monitors, terminals, and motherboards were awaiting repair. India also has the largest density of small shops in the world. The country's misnamed 'unorganized' retail sector, with its bazaars, mandis, and haats, has evolved over centuries into an ecology of 40 million traders, shopkeepers, hawkers, and vendors selling everything one can imagine and a lot more. In the vast informal food sector, fruit and vegetables that go bad are eaten by cows or are composted. Because this amazing retail ecology is labour-intensive, low entropy, low cost, decentralized, and self-organizing, it is also highly efficient – in a word, resilient. In sustainability terms, the unmodernized farming, food, and retail systems of countries like India are state of the art.

If my intuition is correct, and we have already entered a post-growth era, then these small businesses are far better adapted to survive, and even thrive, than their global big brothers – and there are huge numbers of them. Micro-enterprises still constitute the vast majority of businesses in all countries, not just developing ones. One might expect that 75 per cent of people in Africa would work in micro-enterprises – but in Europe, too, more than 90 per cent of businesses are either 'micro', with fewer than ten people, or small, with up to fifty.[13] In the US, anywhere from 70 to 90 per cent of businesses fall into the 'small' category; the vast majority of these are sole proprietorships.[14] And they don't need to grow. Most small businesses are created and run to provide a basic service to their community;

their overheads are for the most part modest; most of their owners are also the workers; and the lion's share of their turnover remains in the local economy.[15] Marius de Geus refers to this global patchwork as a 'utopia of sufficiency'.[16]

MONEY TRAP

The biggest challenge for small firms and informal workers is our dependency on a global money system that must grow in order to survive – even if we do not. What most freelancers, street traders, and small firms value, but usually lack, is the capacity to plan ahead, rather than be perpetually at the mercy of the seasons, or slow-paying clients. We lack liquidity – money – most of the time. We also lack insurance against setbacks and misfortune. Some communities mutualize risk among trusted networks – but they are a minority. Most of the micro-examples I've told you about in this book – not to mention its writer – are therefore in a bind: our micro-economies will never flourish so long as they depend on loans or investment from private banks. These kinds of input bring with them the need to achieve a surplus – a profit – in order to repay the loans with interest. The logic of debt is implacable: it forces an enterprise to turn into a capitalist one, or go bust. This risk is not hypothetical: exactly this process caused the bankruptcy of Mondragón cooperatives in the Basque Country in Spain.[17]

The Open Money Manifesto explains the challenge well:

> The problems with money stem entirely from how conventional money is normally issued. It is created by central banks in limited supply. It's scarce, and hard to get. And we know where it's from: it's from 'them', not 'us'.[18]

It was not always thus. For thousands of years, before it took on a life of its own, money was just a tool. 'The economy' was about livelihoods and the provision of necessities for households. In this context, the local money we used worked, as a system, because we trusted one another as members of a community in a particular place. With the

birth of capitalism, that meaning and purpose of 'economy' was lost
– and money became an end in itself. Speculation for its own sake
replaced provision for family and community – or, as the philosopher
Giorgio Agamben puts it, 'God did not die; he was transformed into
money.'[19]

The system-wide financial crisis of 2008 triggered a plethora of
experiments in alternative money and trading systems that were place-
based and subject to local democratic control. A variety of currencies,
parallel payment systems, and mutual credit schemes were introduced,
and are being tested: Ithaca Hours, Time Dollars, Local Economy
Trading Schemes, Brixton Pounds, micro-credit programmes, interest-
free banking, and other community exchange systems. Some forms
of this so-called solidarity finance have been developing over a longer
period; examples include social banks like Banca Etica Poplare in
Italy, La Nef in France, and JAK bank in Sweden, as well as non-
bank social investment providers like the MAG in Italy and Club
Cigales in France. There is now a crowdfunding website, Goteo, that
is exclusively devoted to raising money for and recruiting volunteers
for commons-based projects. The peer-to-peer technology behind
Bitcoin is the cause of particular excitement at the time of writing;
here, finally, say its advocates, we have a platform for the creation of
money that does not depend on banks. Sadly, the technical approach
of a so-called crypto-currency like Bitcoin contains its own flaws; the
defining technological feature of Bitcoin – a cryptographic invention
called a 'blockchain' – does away with the need for trust among
its users. As the writer Franco Berardi explains, strong social ties are
essential in a healthy money system, and anonymous cryptography
removes the last residues of our social bonds from money, thus
transforming it into the ultimate agent of separation. 'When software
replaces trust,' explains Berardi, 'the last-remaining bit of humanity is
removed from the equation.'[20]

Is money, encrypted or otherwise, unavoidable? Can we feasibly
escape the embrace of a money and debt system that has been
evolving, and constantly reinventing itself, for five thousand years?[21]
By some accounts, we just need to sit and wait; the money economy

cannot persist in its present form, say some, because of its systemic dependency on cheap energy. If less cheap energy is available, as is happening now, the argument goes, and growth in the 'real' economy declines, too, then a money economy that must expand if it is to survive, won't.

A less apocalyptic way to look at the near future is that life without money already exists. In every family and every community on Earth, a lot of work gets done that has never been packaged as jobs, or compensated by payslips. So-called 'non-market' work includes much of the essential activity people have always undertaken to raise and educate their families, take care of their land, and enjoy themselves. Billions of people with low cash incomes meet daily life needs outside the money economy through traditional networks of reciprocity and gifts. They survive, and often prosper, within indigenous social systems based on kinship, sharing, and myriad ways to share resources. As I explained above, this social purpose is one of the reasons so many millions of small enterprises do not need to grow. 'Their goal is working to live,' says Walter Mignolo – 'not living to work.'[22] This autonomy is a strength.

Many of these social practices are very old ones, learned by other societies and in other times. Communitarian relationships with the Earth can be found the world over, from Alaska to Patagonia, and some forms of cooperation and resource sharing date back centuries. These diverse practices of cooperation, mutual aid, reciprocity, and generosity have evolved to meet basic needs in conditions of often extreme resource scarcity: raising children, offering advice or comfort, resolving relational conflicts, teaching basic life skills, cooking, sewing, building the house, farming, raising animals. Although few of these economic activities are integrated within formal economic structures, these parallel systems support a complex ecosystem of small-scale businesses, family farms, local traditions, and extended social and regional networks. Robert Neuwirth, who spent four years among the unregistered markets and roadside kiosks in Lagos, discovered that what happens in this intense world is not simply haphazard; it is a product of intelligence, resilience, self-organization, and group solidarity. The solidarity economy is an economic system – and a dynamic one at that.[23]

The informal economy is incredibly diverse and this is a source of strength. In Africa alone, the variety of different systems of collective self-reliance, and mutual assistance, is stunning; it contains a vast number of ad hoc cooperatives, micro-lending clubs, and group savings and purchasing schemes. My friend Mugendi M'Rithaa, a designer in South Africa, taught me that Africa's parallel economy is based on deeply interpersonal relationships and mutual trust. These voluntary groups are known variously as *stokvels* (cooperative societies) and *boipelogo* (self-reliance) in South Africa; *Bataka kwegaita* (communal solidarity) in Uganda; *nobwa* (reciprocal assistance) in Ghana; *harambee* (pulling together) in Kenya; *ujamaa* (familyhood) in Tanzania; *molaletsa* (collective labour sharing systems) and *motshelo* (group credit unions) in Botswana. Many African cultures, Mugendi told me, are also expert in the arts of communal dialogue, public debate, and consensus-building; they gain additional strength, he told me, from the pivotal role played by women.

Latin America, too, is blessed with diverse forms of solidarity and sharing that are unknown in the North. One of these, described by Walter Mignolo, an Argentine professor, is the *ayllu* – a kind of extended familial community that collectively works a common territory.[24] 'It's akin to the Greek *oikos*, which provides the etymological root for "economy",' Mignolo explains. 'Each *ayllu* is defined by a territory that includes not just a piece of land, but the ecosystem of which that land is one component'; the territory is not considered to be private property but the home of all of those living in and from it. In another commons-based system in Latin America, *marka*, individuals work for one another around the year; one provides labour, the other accommodation and food. The arrangement is reciprocal: after somebody has come to your common to help you work it, you must reciprocate by working on their common. Although the *marka* system dates back to agrarian societies in the Andes, incarnations of the practice are found in modern cities, too. In the *Tequio* system in Mexico, which dates back to pre-Columbian times, community members pool materials and labour to construct schools, wells, and roads. In Brazil, collective mobilizations called *mutirão* harness unpaid work for the construction of community houses where everyone who contributes is a beneficiary. In Mexico's Chihuahua mountains, the word *córima* describes an act of solidarity with someone who's having trouble; it resembles *potlach*, in the Pacific Northwestern United States, where indigenous peoples distributed food and wealth to other tribes who have had a bad season. In the indigenous Amazon region of Colombia and Brazil, a *maloca* is a communal house cohabited by different families; workspaces are shared; at night, the *maloca* becomes a knowledge centre where stories, myths, and legends are told.[25]

BUEN VIVIR

I'm listing these examples at some length in order to persuade you that non-money economies already exist, in abundance, if only one chooses to look. Many of these practices, it is true, are small-scale – but not all of them. The concept of *Buen Vivir*, for example, is an important principle in Ecuador's new (2009) constitution. Rooted in the worldview of the Quechua peoples of the Andes, *sumak kawsay – buen vivir* is its Spanish name – describes a way of organizing daily life that is community-centric, ecologically balanced, and respectful of cultural diversity. Unlike the Western notion of wellbeing, the word 'community' in Buen Vivir includes all living things – not just people. One of its advocates, Eduardo Gudynas, explains that 'Buen Vivir rejects the modern stance that nature is a means to our ends.' Guided by that principle, Buen Vivir campaigners are actively promoting legal and tax reforms, the introduction of environmental accounting, and alternative forms of regional governance. 'Buen Vivir will not stop people building bridges, and it does not reject the use of Western physics and engineering to build them,' explains Gudynas; 'but any bridges we do propose are likely to be placed in locations that serve local and regional needs, and not the needs of global markets.'[26]

Buen Vivir is an inspiring alternative to mainstream models of development – but it has been born into an imperfect world. The ecological promises made by its proponents – principally, progressive

Latin American governments – are not yet being matched in practice. On the contrary: the heavy investments of these governments in health care, education, housing, culture, and social security are being paid for with the proceeds of mining operations and hydrocarbon extraction. Thirty per cent of the world's total investment in mining is in Latin America, and dozens of ecologically damaging open-cut mining projects are under way. In Ecuador, President Rafael Correa argues that his country has no alternative to mining and resource extraction because 'we need this money to end poverty'. His counterpart in Argentina, Cristina Fernández, takes a similar line: 'it is noble to defend flora and fauna, but it's more important to take care of the human species so it has work, water, and sewers'.[27] These contradictory positions have to be faced. Although there will be no jobs, no welfare, and no education on a dead planet, a more positive narrative is needed to counter the toxic allure of extractivism. Although millions of people are busy with projects to meet practical needs in these precarious times, we've been lacking an umbrella concept, a coordinating idea, to make sense of the work we do as individuals in the swarm.

COMMONING

That something – that new story – is the story of the commons. The commons is an idea, and a practice, that generates meaning and hope. In *The Commons: A New Narrative for Our Times*, Silke Helfrich and Jörg Haas talk about the commons as 'all the things that we inherit from past generations that enable our livelihoods'. Seen through that lens, the commons can include land, watersheds, biodiversity, common knowledge, software, skills, or public buildings and spaces. The maintenance, health, and sustainability of these resources are in our shared interest, as they have always been. No individual, company, or government created these common goods; therefore, none has a right to claim them as private property. On the contrary: we inherited them from previous generations and have a moral obligation to look after them for future generations.[28] Along with the other kinds of solidarity I've described in this chapter, the commons as a social practice dates back many centuries. The original meaning of the term comes from the way that communities managed shared land in medieval Europe, but history is filled with similar systems in which communities managed common resources sustainably over the long term. The shared management of water, for example dates back eight thousand years; the earliest records of collectively managed irrigation have been found in regions of the Middle East that we now know as Iraq and Iran. In Bali, as I described in Chapter 4, a complex irrigation society that dates back a thousand years is still alive today, and evolving. The era of globalization is no exception; millions of commoners have organized in recent times to defend their forests and fisheries, reinvent local food systems, organize productive online communities, reclaim public spaces, improve environmental stewardship, and, as David Bollier puts it, to 're-imagine the very meaning of "progress"'.[29]

The commons are not just an ideological proposition. Numerous scientific studies confirm that sharing is in our genes. Psychologists at Harvard, for example, have established that humans possess 'a strong propensity to cooperate rather than compete over limited resources, trusting that they'll benefit in the end'.[30] Swiss scientists, too, have demonstrated that children above a certain age are driven by both genes and social factors to share with others – even if they don't have to.[31]

Stating that the commons are a natural way to organize our world is one thing, but tricky practical questions arise from that: how are we to make sure that the commons are used wisely and fairly? What combination of sticks and carrots is needed in the governance of shared resources? Who should make the rules? No simple formula or rule book exists for commons governance but, in her 1990 book *Governing the Commons*, which was based on a systematic study of fisheries, irrigation, groundwater, and forestry systems, Nobel laureate Elinor Ostrom identified a number of design principles that she had discovered were common in successful examples of self-governance:

▸ the principle that *use value trumps exchange value*: commons that are useful to our everyday life shall not be turned into commodities to be sold for money;

- ▸ the principle of *reciprocity*: anyone who takes from the commons has to contribute to the commons;
- ▸ the principle of *free knowledge*: all commoners must protect the right to share and contribute shared skills and technologies;
- ▸ the principle of *self-organization*: ways to resolve problems are sought for collectively rather than imposed from above.[32]

In the years since Ostrom's pioneering work, a new generation of commoners has added clarity and detail to these broad principles. Among recent additions are:

- ▸ the need for *collaborative monitoring* of biophysical conditions;
- ▸ the principle of *graduated sanctions* to be applied to citizens who violate agreed rules;
- ▸ the need for *conflict-resolution mechanisms*.[33]

For my own work, which often involves connecting designers with communities in transition, I have added a number of Rules of Engagement to complement the principles articulated by Elinor Ostrom and her successors. Among these:

- ▸ *respect what's already there:* most designers are trained to change things first and ask questions afterwards. A better use of a designer's fresh eyes is to reveal hidden value and thus mobilize hidden local resources;
- ▸ *empower local people:* any design action that rearranges places and relationships is an exercise of power. A good test for the sensitivity of a design proposal is whether it enables people to increase control over their own territory and resources.
- ▸ *think whole systems:* when designing an improvement to a common resource, such as a river, the design of the device, such as a pump, will seldom be more than 10 per cent of the complete solution; the other 90 per cent – and the rest of the system – involves distribution, training, maintenance and

service arrangements, and partnership and business models. These are just as important.

An important lesson has emerged from study of these diverse kinds of commoning: the How is as important as the What. Paying attention to the process by which groups work together is just as important as deciding what needs to be done, if not more so. It's not enough simply to proclaim the moral superiority of sharing, for example, and then expect everyone to fall in line. Tough questions must be confronted, and not brushed under the carpet. Among these: how to define, map, and name the resources to be shared; determining who is entitled to what; designing rules and sanctions; designing how to make the rules.

For one long-time advocate of commoning, Massimo De Angelis, how to deal with difference is the most important of these issues by far. 'We have to go beyond the idea that democracy means: "here is my view, there is yours, let's see who wins",' he asserts; 'we need to acknowledge differences, allow those who don't want to share with us, or with whom we do not want to share, to be heard.'[34] Dealing with difference involves a lot of consensus building, active participation, and collective decision-making. All this takes time, and a politics that involves endless meetings is neither attractive nor practicable for most people. New ways of doing politics are therefore needed that are shaped by the ways people live now – not the other way round. For David Bollier, another insightful advocate, commoning is more of an art than a science. 'We all know that the commons is about the stewardship of resources,' he has written, 'but we may not realize that it is also about *hosting* people. Not "managing" them or "organizing" them, but unleashing their capacity to self-organize themselves in creative, constructive, humane ways.'[35] Many commoners cite the free software movement as evidence that such a demanding but effective culture of cooperation is possible. Where there's a will, they say, there's a way.

Another theme among today's new commoners: how to *be* matters just as much as how to meet. Otherwise stated, our inner states are just as important to a healthy political movement as are its external activities. For Sophy Banks, who leads a programme on 'inner transition' for the Transition movement, 'we have a global system that's depleting the planet, and we're burning out ourselves. This is no coincidence. We have to change the culture of how we do things.'[36] For Banks, one indicator of healthy politics is the quality of meetings. In healthy meetings, she observes, people feel relaxed and connected to each other. Even when there's a lot to discuss, time is found to discuss how the group is working, and how people are dealing with differences. The best meetings are celebratory, and not just about building and doing things. A common thread is the need for some kind of caretaker, or steward (or priest, in the case of Bali's water temples) whose job is to nurture communication between members of the community, make sure that everyone understands and abides by the rules, and generally foster a shared spirit of reciprocity and cooperation.

No course exists, as far as I am aware, called commons stewarding or caretaking, but individuals with these special qualities have started to emerge around the world. Cheryl Dahle, founder of the Future of Fish, is one of them.[37] When she began her journey to understand global overfishing, she recalls, she encountered a sprawling and complex tangle of intertwining problems that touched the spheres of policy, commerce, environment, and livelihood. Each of the players in the system has an incredibly personal stake in how we humans choose to rethink the way we hunt, eat, and protect fish. For Dahle, what she names the 'people platform' became the core design challenge: how do you design interactions between competitors – people who've been enemies on opposite sides of policy debate, scientists and business people who don't speak the same language – to enlist their help and collaboration? For insight, Dahle turned to the work of Adam Kahane, one of the pioneers of conflict resolution and collaboration design. 'What might seem at first glance to be abstract theory has proved incredibly instructive,' says Dahle. 'When we convene any group – fishers, processors, or financiers – we set up the conversation so there's something in it for them. We acknowledge the interests of everyone in the room, and we never ask anyone to sacrifice their self-interest. We work to show – to *prove* – that there is a reason for them to shift their thinking and behaviour.'[38]

Another technique with potential for commoning is Appreciative Inquiry (AI). In AI, rather than compile lists of all the problems that need to be fixed, and all the wicked things that have been done, the group focuses first on what's working; it then explores how successful ingredients might be improved, and how. A number of next-generation institutes will teach you skills that are similar to AI. At the Presencing Institute, for example, founder Otto Scharmer runs Theory U workshops that teach people how to 'co-sense and co-create positive change'. The Alia Institute, based in Halifax, Nova Scotia, offers skill-building courses with names like Change Lab and Human Systems Dynamics. Another network, Art of Hosting, teaches people 'how to be successful in complex circumstances when we can't predict what ten, five or even two years down the road will look like'. In Brazil, the Elos Institute, founded in 2000 by young architects, runs a collaborative game called Oasis that's designed 'to awake and give impulse to communities through fast actions with high impact'. A cross between an architectural design project and an Amish-style barn-raising, Oasis games typically end with a square, a park, or a daycare centre being built there and then.

WILD LAW

Throughout this book, I've celebrated all manner of grassroots projects that strike me as examples of positive change. I've also argued that, although mostly small, these myriad activities portend change that is system-wide. Bottom-up gives me hope. But institutional frameworks – especially legal ones – remain vital, too. Laws – and the institutions that impose them – are what people mean by the 'hardwiring' that locks us into damaging relationships with living systems. In most of today's legal systems, for example, only humans have rights. Our laws are based on the Enlightenment notion that the universe is a repository of dead resources for us to exploit, as we choose, for the exclusive benefit of our own species. In contradiction to ecological principles of wholeness and interconnection, property laws divide up land and ecosystems into discrete parcels. Nature's inherent diversity is at odds, too, with free-trade treaties that impose large-scale monoculture projects. Legal systems also underpin farmland grabs by those looking to make money on a capital gain rather than act as stewards of a bioregion.[39]

In *The Great Work*, published in 1999, Thomas Berry called for a new jurisprudence to redefine the relationship between the human community and the Earth community in which it lives. 'We need a legal system that governs the relationship between humans and the natural world as a totality, not as a collection of parts,' wrote Berry, 'we need laws which respect equally the rights of the natural world to exist and thrive.'[40] Is a transformation of our legal systems along these lines remotely feasible? For the South African lawyer Cormac Cullinan, the answer is yes. A pioneer in Earth Jurisprudence, Cullinan compares our situation now with the abolition of slavery: even when American public opinion came to regard slavery as morally abhorrent, he explains, the concept of slaves as property remained hardwired into the legal system. It took a tremendous political effort – not to mention a civil war – before laws were changed and slavery was finally abolished.[41]

Changes to the legal status of living systems are now emerging in a wide variety of legal systems around the world – including unexpected ones. In 1996, for example, a celebrated legal text in the United States called *Should Trees Have Standing?* gave serious consideration to the proposition that trees might be given legal rights – in the same way that minors, or corporations, are given artificial legal personalities.[42] To most people's surprise the Supreme Court, although it voted against the proposal, also found that there was some merit to these arguments. More recently, a dozen US municipalities have introduced ordinances that grant equal rights to human and natural communities. In 2009 the city of Spokane became one of the first cities in the world to legislate for the rights of nature.[43] 'Ecosystems, including but not limited to, all groundwater systems, surface water systems, and aquifers, have the right to exist and flourish,' the measure declared. 'River systems have the right to flow, and to contain water of a quality necessary to provide habitat for native plants and animals, and to provide clean drinking water. Aquifers have the right to sustainable recharge, flow, and water quality.' Then, in 2013, Santa Monica passed a Sustainability Rights Ordinance which recognizes that 'natural communities and ecosystems possess fundamental and inalienable rights to exist and flourish in the City';[44] the Ordinance includes protections for this right from acts by corporate entities which, it states, 'do not enjoy special privileges or powers under the law that subordinate the community's rights to their private interests'. The Ordinance also articulates the rights of people – to self-governance, a healthy environment, and sustainable living. State and Federal authorities have retained the right to veto such measures – but the political lesson is that they are being passed at an increasing rate.

At the scale of the nation state, radical legal expressions of a new worldview are also beginning to emerge. In Latin America, as I mentioned above, Ecuador's national constitution was revised in 2009 to recognize and protect rights of nature.[45] Indigenous elders played a critical part in the revision of the new constitution, which grants to Mother Earth 'the right to exist, persist, maintain, and regenerate its vital cycles, structure, functions, and restoration'. Ecuador's new constitution is not a one-off. In 2010, when Bolivia hosted a World People's Conference on Climate Change and Rights of Mother Earth, it was attended by 30,000 people from 100 countries. One outcome, a Universal Declaration on Rights of Mother Earth, was presented to the UN. And a Global Alliance for the Rights of Nature has been founded with an initial sixty member organizations from around the world. Bolivia itself went on to introduce its own new legislation, an 'Act of the Rights of Mother Earth', and created a new ministry to oversee the Act.[46]

A shift away from seeing Earth solely in terms of 'resources', and the extension of civil rights to the natural world, is beginning to appear in international law and governance, too. An Earth Charter along these lines has been formally recognized by many transnational organizations,[47] and a large number of universities are involved in the

Earth System Governance Project, which was launched in 2009. This multidisciplinary network of scholars and practitioners, working across the Global North and South, is forging new connections between the social and natural sciences in exploring new models of environmental governance.[48]

A NEW CONCEPT OF THE WORLD

This between-two-worlds period of history contains myriad details of an emerging economy in which the word 'development' takes on a profoundly different meaning. Its core value is stewardship, rather than extraction. It is motivated by concern for future generations, not by what 'the economy' needs today. It cherishes qualities found in the natural world, thanks to millions of years of natural evolution. It also respects social practices – some of them very old ones – learned by other societies and in other times. This new kind of development is not backwards looking; it embraces technological innovations, too – but with a different mental model of what they should be used for. With every action we take, however small – each one a new way to feed, shelter, and heal ourselves, in partnership with living systems – the easier it becomes. In the words of Arundhati Roy, 'Another world is not only possible, she is on her way. On a quiet day, I can hear her breathing.'[49] Otherwise stated: we are all emerging economies now.

10 KNOWING: FROM WAYS OF SEEING, TO WAYS OF ACTING

Scientists at Harvard have reported, with great fanfare, that the human mind runs on less energy than a household light bulb.[1] Given its 86 billion neurones, and phenomenal computing power, this is an impressive technical performance. The only shame is that our brain has not proved itself to be as wise as it is energy efficient. In fact, the opposite is the case. Our cool-running brains perceive it as normal to consume non-renewable resources, at an accelerating rate, in a finite world. Even when informed of the grave environmental and social costs of this behaviour, our brains remain unperturbed. In the absence of direct experience to the contrary, they habitually believe that things will turn out for the best.

It's not that our brains lack processing capacity – more that they're processing incomplete data. As I explained in Chapter 1, our whole society has been rendered cognitively blind by a metabolic rift between man and the Earth. Paved surfaces and pervasive media shield us from direct experience of the damage we're inflicting on soils, oceans, and forests. The metabolic rift explains how we're able to put the health of 'the economy' above all other concerns. Its very existence demonstrates that, to repeat Timothy Morton's memorable phrase, 'the ecological catastrophe has already occurred'.[2]

Out of sight, out of mind. Even people whose job it is to think about consequences reason in curious ways. Many intelligent business and money people, for example, are aware intellectually that we live in a finite world – but still believe passionately that growth is a good thing in and of itself. Despite being adepts in a world of numbers, they are curiously unconcerned by the implications of exponential growth. They are also remarkably sanguine about the possibility of unpredictable, non-linear change; they accept intellectually that so-called black swan events can happen – but at some other time, in some other place. Besides, as a senior money guy once told me, that's why they have risk managers. Although some financial actors are indeed using the debt crisis, and austerity, as an excuse to grab publicly owned assets, they are not set on crashing the economy completely. Our situation is even more alarming because these blinkered perceptual frameworks are harder to deal with than openly malign intent.

In this final chapter I discuss three issues, which I believe are related. First, I'll explore the notion that we are living in a 'desert of the real', and what this has meant for environmental campaigns. Second, I will explore alternative ways of knowing about the world – and how these might help us move beyond our present impasse. Finally, and to conclude the book, I turn to the new story that's now emerging about our place in the world – and how this story will bring the change we all yearn for.

DESERT OF THE REAL

Even among people who are supposed to know everything, there's an alarming loss of contact with reality. Consider, for example, the National Security Agency (NSA), whose task is to achieve Total Information Awareness. The machine room for this project, a vast server farm, has been built in the middle of a desert, in Utah, where it needs 11 million litres (3 million US gallons) of fresh water daily just to keep cool. Not terribly smart. The NSA's control room, deep inside the NSA facility, further saps one's confidence in its clarity of thought. A replica of the flight deck of the *Starship Enterprise*, the control room features a solitary commander's seat; this command-and-control model of knowledge is especially ill-suited to the vast volumes of data the centre is allegedly hoovering up.

The NSA's appetite for data is ravenous, but others are not far behind. According to Eric Schmidt, CEO of Google, the amount of data in the world has doubled in the last two years and we apparently create as much information in two days now as we did from the dawn of man until 2003.[3] With so much data available, the argument goes, Big Data represents 'the next frontier for innovation'. Technology firms are promoting the concept of the smart city with particular fervour: their rhetoric imagines the world's conurbations as gigantic train sets that will run so much more smoothly when, by a total coincidence, cities will be spending US$17 billion a year on Big Data back-end services.[4] A craving for data has also enchanted the Quantified Self (QS) movement whose members, adorned with wearable devices, track data on the tiniest details of their physical and psychological status.[5] Whether the sale of 400 million high-end me-meters to otherwise healthy thirty-somethings will ameliorate the pandemic of chronic illnesses in the rest of the population is doubtful – but in Big Data world, quantity counts for more than outcomes.

For the *Financial Times*, Big Data signifies nothing less than the arrival of a 'postmodern economy'. Under the headline 'Welcome to the Desert of the Real', the paper stated in 2012 that 'today's market is the most infinitely complex and impossible object ever imagined'.[6] In order to prosper, the FT opined, the modern investor must be 'adaptable to changing modes of acuity'; be able 'to imagine different realistic states of the world'; and be able to think as 'both the mathematician and the artist'. If frothy prose like this appeared in an undergraduate's cultural studies paper, one would not blink an eye – but these words adorned the house journal of global finance. It is surely alarming that the world's economy is being shaped by people who are mesmerized by all things digital but blind to a much larger reality: the *analogue* knowledge accumulated in nature during 3.5 billion years of evolution.

In his book *Collapse*, Jared Diamond argues that one reason societies fail is that their elites are insulated from the negative impact of their own actions.[1] Diamond focuses on Easter Island, where the overuse of wood products eventually destroyed its inhabitants' survival prospects, but the lesson applies equally to us today. We lust for speed, perfection, and control but, because we inhabit an abstract, digitally diminished world, we're blind to the true costs of what we wish for. I do not pretend to be a cognitively superior observer, here; I spend too much time myself in environments, such as airport lounges, that are just as insulated from reality as the FT's news room or a risk trader's console. But I also spend enough time outside the digital bubble to know that the environmental impacts of the economy are no less devastating just because they are out of sight.

The desert of the real isolates from literally vital knowledge in four ways: because it's invisible; because it's somewhere else; because our sensory bandwidth is too narrow; and because we're 'educated'.

Of the life-critical phenomena we don't see because they're invisible, the most important is energy intensity. As I explained in Chapter 1, we need sixty times more energy per person to meet our daily life 'needs' than pre-modern men and women – and that gap is widening. When you think about it, that sixty-fold-and-rising difference *should* be terrifying – but we don't think about it, or not clearly. Then there are life-critical developments that we ignore because they're happening somewhere else. Our economy's ravenous appetite for external nutrient supplies is a case in point; although these flows have grown 1,500 times in just 50 years – an astounding rate of increase – their environmental and social costs hardly disturb us at all. Why? Mainly because these costs are being paid by other people, somewhere else. The toxic rivers of slurry produced during mining the rare metals that are used in all of our cellphones? They don't touch us directly, so we don't think about them.

Another deadly feature of the desert of the real is that we think too much, and sense too little. Think back to that brain and its billions of neurons. We only use a tiny fraction of those neurons for conscious observation and rational thought: we use the rest to experience the world *unconsciously* – but these other ways of knowing the world have been suppressed in modern society. For the philosopher John Zerzan, this is where our problems began – when we embraced symbolic culture and placed language, art, and number above other ways of knowing the world. Because every abstraction both simplifies, and distances, earthly reality, it underpins a concept of progress in which the globe is perceived to be a repository of resources to fuel endless growth.[8]

Abstract thought is deeply embedded in the fourth defining feature of the desert of the real: the fact that we've been 'educated'. Time was, not so long ago, when children didn't go to school: school surrounded them. As Ellen Haas describes it, 'nature was a living teacher. Every relative – and every plant and animal – was a mentor. People soaked up the language of plants and animals by immersion.'[9] We are born with an inherited aesthetic tendency to appreciate this intimate connection with the world, and for thousands of years this form of learning served us well – but nowadays we go to school. There, an unremitting focus on science and technology exacerbates our dislocation from the Earth. We become expert in the manipulation of symbols, abstractions, and concepts – but to what end? To earn money? To consume? We are also mistaught in school that competition between individuals, and survival of the fittest, is the dominant framework of life on Earth. The more up-to-date theory of symbiogenesis suggests that evolution takes place in communities of interacting entities and that, as Donna Haraway puts it, 'our environment is us'.[10] School-based education separates children physically, and therefore cognitively, from this reality. By promulgating a way of knowing that assumes an external world of objects and facts, it invalidates local systems of knowledge and local ways of knowing as well.[11]

ENVIRONMENTAL COMMUNICATIONS

For forty years, environmental campaigns have floundered in this desert of the real. It's not as if scientists, designers, and artists have been idle while the biosphere suffers. The green movement has been enriched by a dazzling array of maps, images, data sets, and visualizations. Much of this creative work is impressive, even striking

– but an unrelenting flow of gloomy news, on its own, has proved at best ineffective, at worst, counterproductive. 'Doomer porn', as some call it, stubbornly resists empathy. It produces guilt and denial rather than transformational change.

As an alternative to doomer porn, more recent campaigns to change our behaviour have accentuated the positive and feature real people helping to make the world healthier. The aim is to make green behaviour more normal, more 'sticky'. I'm sympathetic to this intent – as a writer I, too, try to focus on the positive – but campaigns to 'raise awareness' suffer from a structural flaw. However positive and uplifting their stories may be, they leave untouched the underlying narrative that we can have our cake and eat it – where 'cake' means a perpetual growth economy. To be blunt: a focus on the individual's personal contribution to a problem – and how to change that – is an example of what cynical politicians call bait and switch. Two simple examples: if you or I take our shopping home in a reused disposable plastic bag, and feel good about doing so, the bag is typically responsible for about one-thousandth of the footprint of the food it contains.[12] Or if I turn off my phone charger, on the principle that every little helps, the energy saved in switching off for one day is used up in *one second* of driving a car.[13] Reusing a bag or turning off a charger may be an easier thing to ask than something huge and abstract, like reshaping a food system – but campaigns to make us feel good about ourselves deflect attention from the underlying values and structures that shape our behaviour in the first place.

If accentuating the positive is not, of itself, an answer, we are left with a dilemma: what are we to do if, when people are exposed to shocking stories and images, nothing seems to change in the system as a whole? What are we to do as designers if we create a powerful piece of communication – and it has no impact? How do we reach a TL;DR generation (textspeak for Too Long: Didn't Read) that survives the media blitz by filtering most of it out? In my search for guidance on this topic, I've discovered these are not new questions. St Augustine, in *City of God*, attacked 'scenic games' as being responsible for the death of the soul – and that was more than

1,500 years ago. A century ago, in 1908, the American philosopher John Dewey decried the emergence of what he called a 'Kodak fixation'[14] – a photographic attitude that reduces the citizen's role to that of a spectator, detached from that which is experienced. Ivan Illich, writing in 1971, believed that our culture started to go off the rails when monks stopped reading texts aloud to each other and became solitary scholars – in 1120.[15] Twenty years ago, Susan Sontag's classic text *Regarding the Pain of Others* raised similar issues – with particular reference to war photography.[16] 'Why is it', she asked, 'that even when we are exposed to shocking stories and images, nothing seems to change?' Sontag memorably alerted us to the danger that photographs – and by implication all visualizations – have a tendency, in her words, to 'shrivel sympathy'. Images shown on television, she wrote, are, by definition, images of which one sooner or later tires. Image-glut keeps attention light, mobile, relatively indifferent to content. 'Compassion', Sontag concluded, 'is an unstable emotion… it needs to be translated into action, or it withers…. It is *passivity* that dulls feeling.'

If it's passivity that dulls feeling, as Susan Sontag concluded, if emitting messages – however clever or evocative they may be – is ineffective without some kind of follow-up action, then it follows that the actions we need to take are those that reconnect us – viscerally and emotionally – with the living systems we've lost touch with. These actions should create space for people to experience relationships with living systems no matter how small the scale; they should facilitate a sense of belonging and being at home in the world as it is now, in Roger Scruton's words, and focus attention on the positive qualities of often small, humble, living things that surround us.[17]

NEW WAYS OF KNOWING

In today's world we learn to perceive the things around us as stuff that is lifeless, brute, and inert. Nature, if we think about it at all, is a nice place to go for a picnic. With this picture of the world in mind, we fill up our lives, lands, and oceans with junk without a second thought. But we used to think quite differently. The idea that things might be 'vital' was first expounded formally by Greek philosophers known as 'hylozoists' – 'those who think that matter is alive'; they made no distinction between animate and inanimate, spirit and matter. For Roman sages, likewise: in his epic work *On the Nature of Things*, the poet Lucretius argued that everything is connected, deep down, in a world of matter and energy. Ancient Chinese philosophers also believed that the ultimate reality of the world is intrinsically dynamic; in the *Tao*, everything in the universe, whether animate or inanimate, is embedded in the continuous flow and change. In Buddhist texts, images of 'stream' and 'flow' appear repeatedly; they evoke a universe that's in a state of impermanence, of ceaseless movement. In seventeenth-century Europe, the Dutch philosopher Baruch Spinoza conceived of existence as a continuum, an inseparable tangle of body, mind, ideas, and matter. And just seventy years ago, Maurice Merleau-Ponty was an advocate not only of being in the world but also of belonging to it, having a relationship with it, interacting with it, perceiving it in all dimensions.

Our long-held belief that matter matters, so to speak, was obscured by the fire and smoke of the thermo-industrial economy – but it never really went away, and many of these ideas are resurfacing today. For thinkers in the 'new materialism' movement, our relationship with the material world would be more respectful and joyful, if only we realized that we are part of the world of things, not separate from it.[18] Timothy Morton, for example, is adamant that there is more to 'things' than we know in the 'vast, sprawling mesh of interconnection without a definite centre or edge' that constitutes our world.[19] Another philosopher, Jane Bennett – responding, in her words, to a 'call from our garbage' – advocates a patient, sensory attentiveness to what she calls the 'vibrancy' of matter and the non-human forces that operate outside and inside the human body.[20] Our wasteful patterns of consumption would soon change, she reckons, if we saw, heard, smelled, tasted, and felt all this litter, rubbish, and trash as 'lively' – not just inert stuff. 'Sometimes those sticking their heads in the sand are looking for something deep,' quips yet another philosopher, Peter Gratton; when everything around is understood to be 'vital', he asks, what political and ethical consequences follow? Do bacteria count as life? Viruses? A robot? Is the ecosystem itself a life? If the answer to any of these questions is yes, or even maybe, Gratton argues, then the assumption that we humans have a right to exploit the world to our own ends begins to break down.[21]

CONNECTED

Philosophers have argued about the difference between humans and things for more than two thousand years. They'd probably carry for another two thousand except that developments in science seem poised to settle the argument once and for all. Since the 1980s, scientific discoveries have confirmed the proposition that no organism is truly autonomous. In Gaia theory, systems thinking, and resilience science, researchers have shown that our planet is a web of interdependent ecosystems; the dead, mechanical object that has shaped scientific thought for most of the modern age turns out to have been misguided. The study of everything from sub-microscopic viruses, yeasts, ants, mosses, lichen, slime moulds, and mycorrhizae, to trees, rivers, and climate systems, adds up to a new story. Not only are all natural phenomena connected, but their very essence is to be in relationships with other things – including us. On a molecular, atomic, and viral level, humanity and 'the environment' literally merge with one another, forging biological alliances as a matter of course.[22]

As I explained earlier, our education does not equip us to grasp these hidden connections, but for thinkers such as Fritjof Capra, this knowledge is literally vital. The greatest challenge of our time, he believes, is to foster widespread awareness of the hidden connections among living and non-living things.[23] In a powerful response to Capra's challenge, Stephan Harding, in his book *Animate Earth*, describes how the world works not only at the macro level – the atmosphere, oceans, or Earth's crust – but also on a micro level: the way plankton and bacteria contribute to the formation of clouds (by acting as nuclei for water droplets); how mycorrhizal fungi team up with plants that grow in poor soils; how chemical signals called pheromones allow ant colonies to behave like a super-organism.

Co-evolution – the formation of biocultural partnerships – is how our fertile planet thrives, says Harding. Although we have ruptured these relationships, it is not too late to build bridges so that Earth can become healthy and self-regulating once again.[24]

These scientific findings resolve a question that has vexed philosophers more than any other: where does the mind end, and the world begin? Until recently, we tended to think of the nervous system as a glorified set of message cables connecting the body to the brain – but from a scientific perspective, the boundary between mind and world turns out to be a porous one. The human mind is hormonal, as well as neural. Our thoughts and experiences are not limited to brain activity in the skull, nor are they enclosed by the skin. Our metabolism, and nature's, are interconnected on a molecular, atomic, and viral level. Mental phenomena – our thoughts – emerge not merely from brain activity, but from what Teed Rockwell describes as 'a single unified system embracing the nervous system, body, and environment'.[25] The importance of this new perspective is profound. If our minds are shaped by our physical environments – and not just by synapses clicking away inside our box-like skulls – then the division between the thinking self and the natural world – a division that underpins the whole of modern thought – begins to dissolve. Having worked hard throughout the modern era to lift ourselves 'above' nature, we are now being told by modern science that man and nature are one, after all.

THE SAVAGE MIND

This is something that 'savage' people have known all along. From Alaska to Patagonia, and over millennia, indigenous peoples have had a closer connection with the biosphere than we do. In hundreds of cultures, the belief that non-human entities possess vital or spiritual qualities is commonplace. All manner of ceremonies, arts, participatory ritual, and performance, are used to maintain harmony with animals, plants, objects, and nature. This is not to suggest that indigenous people live in a bubble – still less that they should be excluded from the modern world as museum pieces. But for people who enjoy embodied connection with living systems on a daily basis, the idea of co-evolution with all living things is – well, second nature.

Animism – the belief that there is no separation between the spiritual and physical or material world – is not confined to pre-modern cultures. As I touched on above, many vibrant world religions – among them Shinto,[26] Serer, Hinduism, Buddhism, and Jainism – contain elements of a belief that plants, rocks, mountains, and even thunder, wind, or shadows, are fellow sentient subjects. Plants and fungi are especially revered in Amazonian cultures; for 'plant teachers' or 'vegetalista', the medicinal and magical properties of plants are a means to see deeper into the spiritual ecology of deep forests.[27] To those who live with it, this perspective is not at all exotic; most indigenous people do not even have a word to describe their matter-of-fact belief. The word 'animism' was invented by Western anthropologists.

Traditional cultures may have fewer abstract theories and concepts than the modern world, but they experience the living world more richly than we do. A forest-dwelling honey gatherer, for example, needs to be very still and attentive to pick up the faint and distant hum of bees; a desk-bound writer, preoccupied by ideas and deadlines, would miss these signals. And although indigenous cultures don't have schools like we do, transrational perception is cultivated. Altered states of consciousness – meditation, trance, dreams, and imagination – are found in 90 per cent of traditional cultures. The anthropologist Tara Waters Lumpkin, who has immersed herself in the subject first-hand, believes that perceptually diverse cultures are better equipped to practise whole-systems thinking than people trained only in the scientific method; and, because they experience biodiversity in richer and more complex ways, they are better stewards of their environments. Perceptual diversity leads to a higher degree of adaptability and evolutionary competence, Lumpkin concludes, than the monophasic consciousness, embodied in the scientific approach, that is disconnected from richer modes of understanding.[28]

The Western tendency to represent nature with abstractions is a challenge – but it's not an either/or choice. Scientific and indigenous knowledge should complement each other; we need to learn how to navigate freely among a diverse ecology of information actors and resources. Luckily, we probably still have the aptitude. Having spent 99 per cent of our social history in hunting and gathering environments, our sensitivity to landscape is genetically hardwired: when to move, where to settle, which activities to follow in various localities. Flowers, sunsets, clouds, thunder, snakes, and predators: all these are environmental signals that trigger programmed-in response systems. The anthropologist Claude Lévi-Strauss is especially reassuring about our species; 'we all have within us the call of the wild,' he wrote. 'The savage mind is our mind.'[29]

BEING THERE

In 1909, Peter Kropotkin was asked whether it was possible to learn a trade as difficult as gardening from books. 'Yes, it is possible,' he replied, 'but a necessary condition of success, in work on the land, is communicativeness – continual friendly intercourse with your neighbours.' Although a book or a classroom lecture can offer good general advice, Kropotkin explained, every acre of land is unique. Each plot is shaped by the soil, its topography and biodiversity, the wind and water systems of the locality, and so on. 'Growing in these unique circumstances can only be learned by local residents over many seasons,' the aristocratic anarchist concluded. 'The knowledge which has developed in a given locality, that is necessary for survival, is the result of *collective* experience.'[30]

One reason we've damaged our own life places is that we under-value the kinds of socially created knowledge Kropotkin wrote about. Embodied, situated, and unmediated communications were the norm before we invented writing and, later, mass media. It follows that, at a social level, we need to talk to each other more – and face to face. The philosopher and theologian Martin Buber counselled just such activity in his book *I and Thou*[31] in 1923. 'All knowledge is dialogic', wrote Buber – but he did not just advocate talk. Connection is not just about words, he taught, it's about encounter and community. Literally 'vital' conversations need to be embodied, and situated. It follows from Martin Buber's insights that we need more interactive and less choreographed forms of encounter. Over many thousands of years participatory ritual, and performance, were the main ways in which beliefs were shared within a culture. In indigenous cultures the world over today, too, communities use ceremonies, arts, and stories to maintain harmony between nature and culture, body and mind. The meeting formats we design now, therefore, should enable us, quite simply, to breathe the same air in a natural context. In my own work as an event organizer I call these 'feral encounters': they usually takes place outside – or at least, outside the disciplinary tent – and are shaped and energized by their context, not by an abstract agenda. Being outside the tent also brings one closer to people with first-hand experience of social-ecological systems: fishers, farmers, foresters, water stewards, ride-sharers, space reusers. An out-of-the-tent approach enables distracted people to cherish and nurture what's unique about each place, each moment, each group of people. This is why we call our Doors of Perception encounters 'xskools'. 'X' means: breathing the same air. Shoulder-to-shoulder learning. The opportunity to be still. Only here, only now.

HOW CHANGE HAPPENS

Getting out of the tent is just the start. Whether inside the tent or out, change doesn't happen just because you tell people things. Change is not about campaigns to raise awareness, or to change other people's behaviour; these approaches simply don't work – or only partially. Change is not much about finely crafted 'visions' and the promise of a better reality in some future place and time. In the fine words of Guy McPherson, 'Nature evolves by paying attention to present needs and opportunities – working piecemeal rather than in grand designs.'[32] Above all, change is not about making demands, of telling politicians what they *must* do. 'The government must end our dependency on fossil fuels.' 'We must end this obsession with perpetual growth.' 'They' won't do any such thing. They can't. They're not captains of a ship. They're following the wrong script in a dysfunctional system.

Change is more likely to happen when people reconnect –
with each other, and with the biosphere – in rich, real-world contexts
of the kind I have written about in this book. This will strike some
readers as being naive and unrealistic. But given what we know
about the ways complex systems – including belief systems – change,
my confidence in the power of the Small to shape the Big remains
undimmed. As we've learned from systems thinking, transformation
can unfold quietly as a variety of changes and interventions, and
often small disruptions accumulate across time. At a certain moment
– which is impossible to predict – a tipping point, or phase shift, is
reached and the system as a whole transforms. It's a lesson confirmed
repeatedly by history: 'All the great transformations have been
unthinkable until they actually came to pass,' writes the French
philosopher Edgar Morin. 'The fact that a belief system is deeply
rooted does not mean it cannot change.'[33]

OUR ANIMATE WORLD

I hope I have persuaded you, in this short book, that profound change
is already underway – and not just at the level of exotic ideas. Across
the world, a multitude of social movements and grassroots projects are
also animated by the recognition that our lives are codependent with
plants, animals, air, water, and soils. These are the green shoots of
a leave-things-better economy – myriad projects in which people are
taking action to close the metabolic rift.

The philosopher Joanna Macy describes the appearance of this
new story as the 'Great Turning', a profound shift in our perception,
a reawakening to the fact that we are not separate or apart from
plants, animals, air, water, and the soils. There's a spiritual dimension
to this story – Macy is a Buddhist scholar – but her Great Turning
is consistent with recent scientific discoveries, too: the idea, as
articulated by Stephan Harding, that the world is 'far more animate
than we ever dared suppose'. Explained in this way – by science,
as much as by poetry, art, and philosophy – the Earth no longer
appears to us as a repository of inert resources. On the contrary: the
interdependence between healthy soils, living systems, and the ways

we can help them regenerate, finally addresses the 'why' of economic activity that we've been lacking. This narrative points to the one kind of growth that makes sense, and that we can afford: the regeneration of life on Earth. The core value of this emerging economy is stewardship, rather than extraction – and the positive feeds on the positive. The more pieces we fit in – each piece a new way to feed, shelter, and heal ourselves in partnership with living processes – the easier it becomes. It's our genes at work: formed long before the industrial age, they're helping to reconnect us with our wild side.

NOTES

CHAPTER I: CHANGING

1 Vidyut, 'Smart Cities or Cleverly Disguised Corporate Colonies?', Aam janata, 14 February 2015, https://aamjanata.com/smart-cities-cleverly-disguised-corporate-colonies/

2 Bardi, Ugo, 'Tainter's law: Where is the physics?', Our Finite World, 27 March 2011, http://ourfiniteworld.com/2011/03/31/tainters-law-where-is-the-physics/

3 Fernández-Savater, Amador, 'Strength and Power Reimagining Revolution', Guerilla Translation, 29 July 2013, http://guerrillatranslation.com/2013/07/29/strength-and-power-reimagining-revolution/comment-page-1/

4 Macy, Joanna, 'The Great Turning', Ecoliteracy, http://www.ecoliteracy.org/essays/great-turning

5 Morin, Edgar, *Homeland Earth: A Manifesto for the New Millennium – Advances in Systems Theory, Complexity and the Human Sciences*, New York: Hampton Press, 1999

6 DeLong, Brad, 'Earl Cook's Estimates of Energy Capture', Grasping Reality, 22 January 2012, http://delong.typepad.com/sdj/2012/01/earl-cooks-estimates-of-energy-capture.html

7 Murphy, Tom, 'Can Economic Growth Last?', Do The Math, 14 July 2011, http://physics.ucsd.edu/do-the-math/2011/07/can-economic-growth-last/

8 Tverberg, Gail, 'Energy and the Economy – Twelve Basic Principles', Our Finite World, 14 August 2014, http://ourfiniteworld.com/2014/08/14/energy-and-the-economy-twelve-basic-principles/

9 Glover, John, 'Global Debt Exceeds $100 Trillion as Governments Binge', Bloomberg, 9 March 2014 http://www.bloomberg.com/news/2014-03-09/global-debt-exceeds-100-trillion-as-governments-binge-bis-says.html

10 Tverberg, Gail, 'WSJ Gets it Wrong on "Why Peak Oil Predictions Haven't Come True"', Our Finite World, 6 October 2014, http://ourfiniteworld.com/2014/10/06/wsj-gets-it-wrong-on-why-peak-oil-predictions-havent-come-true/

11 Commons Strategy Group, 'The Coming Financial Enclosure of the Commons', Shareable, 11 June 2013, http://www.shareable.net/blog/the-coming-financial-enclosure-of-the-commons

12 'The Financialisation of Nature: Linking food, land grabs, climate & mining', Gaia Foundation, 10 November 2011, http://www.gaiafoundation.org/blog/the-financialisation-of-nature-linking-food-land-grabs-climate-mining

13 http://naturenotforsale.org/declaration/

14 https://www.foeeurope.org/nature-not-for-sale

15 http://www.weforum.org/reports/global-risks-report-2015

16 http://www.dni.gov/index.php/about/organization/national-intelligence-council-global-trends

17 'Planetary Boundaries', Stockholm Resilience Centre, http://www.stockholmresilience.org/21/research/research-news/1-15-2015-planetary-boundaries-2.0---new-and-improved.html

18 Raford, Noah, 'Collapse Dynamics: Phase Transitions in Complex Social Systems', 29 November 2009, http://news.noahraford.com/?p=48

19 Kelly, Kevin, 'The Post-Productive Economy', The Technium, 1 January 2013, http://kk.org/thetechnium/2013/01/the-post-produc/

20 http://www.weforum.org/events/world-economic-forum-annual-meeting-2013

21 Murphy, Tom, 'Can Economic Growth Last?', Do The Math, 14 July 2011, http://physics.ucsd.edu/do-the-math/2011/07/can-economic-growth-last/

22 Zolli, Andrew and Ann Marie Healy, *Resilience: Why Things Bounce Back*, New York: Simon & Schuster, 2013

23 Foster, John Bellamy, *The Ecological Revolution: Making Peace with the Planet*, New York: Monthly Review Press, 2009, p. 13

24 Morton, Timothy, 'The Catastrophe Has Already Occurred', 13 July 2008, http://ecologywithoutnature.blogspot.fr/2008/07/catastrophe-has-already-occurred.html

25 Kuhn, Thomas S., *The Structure of Scientific*

Revolutions, University of Chicago Press, 1962
26 'World in Transition: A Social Contract for Sustainability', German Advisory Council on Global Change (WGBU), http://www.wbgu.de/fileadmin/templates/dateien/veroeffentlichungen/hauptgutachten/jg2011/wbgu_jg2011_en.pdf
27 Kingston, Christopher and Gonzalo Caballero, 'Comparing Theories of Institutional Change', Amherst College, 16 June 2008, https://www3.amherst.edu/~cgkingston/Comparing.pdf

CHAPTER 2: GROUNDING
1 'Les Incroyables Comestibles du tarn au Festival Cinéfeuilles à Gaillac', Incroyables Comestibles Castres, 14 November 2013, http://incroyablescomestiblescastres.blogspot.fr/2013/11/droit-aux-buttes-les-avantages-et-les.html
2 'The Implications of Ecological Restoration on Microarthropod Diversity', 10 Things Wrong With Environmental Thought, 30 October 2012, http://10thingswrongwithenvironmentalthought.blogspot.fr/2012/10/why-are-implications-of-ecological.html
3 Stamets, Paul, *Mycelium Running: How Mushrooms Can Help Save the World*, Berkeley: Ten Speed Press, 2005
4 http://www.summerofsoil.se/forum/
5 Merryweather, James, 'Secrets of the Soil', *Resurgence*, No. 235, March/April 2006, pp. 26–8
6 'Land Use Reduction', Umweltbundesamt, 12 May 2014, http://www.umweltbundesamt.de/en/topics/soil-agriculture/land-use-reduction
7 Pappas, S., 'Vanishing Forests: New map details global deforestation', *Live Science*, 16 November 2013, http://www.livescience.com/41215-map-reveals-global-deforestation.html
8 Bossio, Deborah, 'A month on Land: Restoring soils and landscapes', CGIAR, 18 October 2013, http://wle.cgiar.org/blogs/2013/10/18/a-month-on-land-restoring-soils-and-landscapes/
9 Leavergirl, 'Mother Nature's Tillers', Leaving Babylon, 18 October 2014, http://leavingbabylon.wordpress.com/2014/10/18/mother-natures-tillers/#comment-8573
10 Withnall, Adam, 'Britain has only 100 harvests left in its farm soil', *Independent*, 20 October 2014, http://www.independent.co.uk/news/uk/home-news/britain-facing-agricultural-crisis-as-scientists-warn-there-are-only-100-harvests-left-in-our-farm-soil-9806353.html
11 Jones, Christine, 'How to Build New Topsoil', Managing Wholes, 8 February 2010, http://managingwholes.com/new-topsoil.htm
12 http://www.doorsofperception.com/development-design/whole-whole-on-the-range/
13 'Organic No Till', Rodale Institute, 11 February 2015, http://rodaleinstitute.org/our-work/organic-no-till/
14 Drescher, James W., 'Enrichment Forestry at Windhorse Farm', November 2013, http://www.windhorsefarm.org/wp-content/uploads/2013/11/Enrichment_Forestry.pdf
15 Olsson, P., L. H. Gunderson, S. R. Carpenter, P. Ryan, L. Lebel, C. Folke, and C. S. Holling, 'Shooting the rapids: Navigating transitions to adaptive governance of social-ecological systems', *Ecology and Society*, Vol. 11, Issue 1, 2006, http://www.ecologyandsociety.org/vol11/iss1/art18/, http://www.ucpress.edu/book.php?isbn=9780520236288
16 Thayer, Robert L. Jnr, *LifePlace: Bioregional Thought and Practice*, University of California Press, 2003
17 Viganò, Paolo, 'Lezing: Paola Viganò – The Horizontal Metropolis', Universiteit Gent, April 2012, http://www.architectuur.ugent.be/2012/04/lezing-paola-vigano-the-horizontal-metropolis/
18 Lyle, John Tillman, *Regenerative Design for Sustainable Development*, London: Wiley, 1996
19 http://www.landstewardship.eu/your-role/citizens
20 http://www.rightlivelihood.org/karaca.html
21 http://www.hedgelink.org.uk/
22 Navntoft, Søren et al., 'Buffer Zones for Biodiversity of Plants and Arthropods: Is there a compromise on width?', Pesticides

Research No. 127, Danish Ministry of the
Environment, 2009, http://www2.mst.dk/
udgiv/publications/2009/978-87-92617-09-5/
pdf/978-87-92617-10-1.pdf
23 http://www.centreforstewardship.org.uk
24 http://peoplefoodandnature.org/about-
integrated-landscape-management
25 Friedman, Rachel and Seth Shames,
'Recognizing Common Ground: Finding
meaning in integrated landscape
management', *Landscapes for People, Food
and Nature*, 13 November 2013, http://
beta.landscapes.ecoagriculture.org/
blog/recognizing-common-ground-
finding-meaning-in-integrated-landscape-
management/
26 http://bioregion.evergreen.edu/aboutus.
html
27 http://landscapes.ecoagriculture.org
28 McIntosh, Alastair, *Soil and Soul: People
versus Corporate Power*, London: Aurum Press,
2001

CHAPTER 3: WATERKEEPING
1 http://wwf.panda.org/what_we_do/
footprint/water/dams_initiative/;
see also Barlow, Maude, 'Water Abuse and
Climate Change', Our Water Commons, 11
February 2015, http://ourwatercommons.org/
feature/water-abuse-and-climate-change
2 Hoekstra, A. Y. (ed.), 'Virtual Water Trade',
Value of Water Research Report Series No. 12,
2003, UNESCO-IHE, Delft
3 Aldaya, M. M. and A. Y. Hoekstra, 'The
Water Needed for Italians to Eat Pasta and
Pizza', Waterfootprint, May 2009, http://
www.waterfootprint.org/Reports/Report36-
WaterFootprint-Pasta-Pizza.pdf
4 Angelakis, Andreas N. et al., *Evolution
of Water Supply Through the Millennia*,
International Water Association, 2012
5 Thackara, John, 'From King
Parakramabahu to Ethical Fashion', Doors of
Perception, 3 December 2009, http://www.
doorsofperception.com/development-design/
from-king-parakramabahu-to-ethical-fashion/
6 Elvidge, Christopher D., et al., 'Global
Distribution and Density of Constructed
Impervious Surfaces', *Sensors*, 21 September

2007, http://www.mdpi.com/1424-
8220/7/9/1962
7 Hayward, Keith (ed.), 'A World Of
Opportunities', International Water Association,
2011, http://www.iwapublishing.com/pdf/
WorldofOpportunities_2011-2012.pdf
8 http://www.watershedmg.org/
9 Baradwaj, Aajwanthi, 'Towards a Water-
Sensitive City: The story of Jakkur Lake',
Bangalore Citizen Matters, 12 February 2014,
http://bangalore.citizenmatters.in/articles/
en-route-to-a-water-sensitive-city-the-story-
of-jakkur-lake?utm_source=ref_article
10 'The Ugly Indian', https://www.facebook.
com/theugl.yindian
11 Cave, Damien, 'A Vision of an Oasis
Beneath the Sprawl', *New York Times*, 31 May
2011, http://www.nytimes.com/2011/06/01/
world/americas/01mexico.html?_r=0
12 Taller 13 Arquitectura, 'ITDP_
RegeneracionLaPiedad', 2011, http://issuu.
com/taller13/docs/itdp_regeneracionlapiedad
13 'Seoul, South Korea: Cheonggye
Freeway', Preservation Institute, 2007,
http://www.preservenet.com/freeways/
FreewaysCheonggye.html
14 'London Rivers Action Plan', River
Restoration Centre, January 2009,
http://www.therrc.co.uk/lrap/lplan.pdf
15 Jowit, Juliette, 'River Rescue: Project
launched to breathe life into waterways
buried under London concrete and brick',
The Guardian, 8 January 2009, http://www.
theguardian.com/environment/2009/jan/08/
river-restoration-london
16 Saunders, Peter, 'Purple Root Water
Hyacinth A Natural Remedy for Pollution',
Institute of Science in Society, 9 September
2013, http://www.i-sis.org.uk/Water_
Hyacinth_A_Remedy_for_Pollution.php
17 Smith, Jeff, 'Sweet Solution? Licorice
could reclaim degraded lands', CGIAR,
29 October 2013, http://wle.cgiar.org/
blogs/2013/10/29/sweet-solution-licorice-
could-reclaim-degraded-lands/
18 http://www.britishflora.co.uk/
19 'One Million Cisterns for Water
Harvesting in North East Brazil', DryNet, 9
September 2014, http://www.dry-net.org/

index.php?page=3&successsstoryId=21

20 Quintana, J. et al., 'PROJETO SERTÃO: Applying the agro-ecological approach to ensure sustainability in the semi-arid Brazilian northeast', The Green Line, November 2009, http://www.thegef.org/gef/greenline/Nov09/Partner1.html

21 Silverman, Howard, 'Resilience & Transformation: A Regional Approach', Ecotrust, 30 January 2012, http://www.ecotrust.org/media/Resilience_Report_013012.pdf

22 Elbakidze, M., P. K. Angelstam, C. Sandström, and R. Axelsson, 'Multi-Stakeholder Collaboration in Russian and Swedish Model Forest Initiatives: Adaptive governance toward sustainable forest management?', *Ecology and Society*, Vol. 15, Issue 2, 2010, http://www.ecologyandsociety.org/vol15/iss2/art14/

23 Venkata, Samala, Govardhan Das, and Jacob Burke, 'Smallholders and Sustainable Wells: Participatory groundwater management in Andhra Pradesh', Food and Agriculture Organization, 2013, http://www.fao.org/docrep/018/i3320e/i3320e.pdf

24 Thackara, John, 'Venice: From Gated Lagoon to Bioregion', Doors of Perception, 2 December 2012, http://www.doorsofperception.com/sustainability-design/venice-from-gated-lagoon-to-bioregion/

25 Bergsten, Arvid et al., 'Connecting Landscapes', Stockholm Resilience Centre, 18 February 2015, http://www.stockholmresilience.org/21/research/research-news/2-18-2015-fit-to-work.html?utm_source=Stockholm+Resilience+Centre+newsletter&utm_campaign=0aa574179a-Newsletter_September_20148_26_2014&utm_medium=email&utm_term=0_216dc1ed23-0aa574179a-95166069

26 Thackara, John, 'Why Bill Gates Needs to Listen to More Gamelan Music', *Design Observer*, 18 July 2012, http://observatory.designobserver.com/johnthackara/feature/why-bill-gates-needs-to-listen-to-more-gamelan-music/35218/

27 Erviani, Ni Komang, '"Subak" Inscribed on UNESCO World Heritage List', *Bali Daily*, 3 July 2012, http://www.thejakartapost.com/bali-daily/2012-07-03/subak-inscribed-unesco-world-heritage-list.html

28 Becker, Judith, 'Time and Tune in Java', in A. L. Becker and A. Yengoyan (eds), *The Imagination of Reality: Essays in Southeast Asian Coherence Systems*, Norwood, NJ: Ablex Publishing Co., 1979

29 Vismanath, S., 'Fostering a Water Culture', Rainwater Harvesting, 13 July 2013, http://rainwaterharvesting.wordpress.com/2013/07/13/on-the-naxi-water-culture/

30 Laureano, Pietro, *The Water Atlas: Traditional Knowledge to Combat Desertification*, Unesco/Bollati Boringhieri editore s.r.l., 2001, may be downloaded from https://www.box.com/s/xroo4hsq2bl190vjuu4x

CHAPTER 4: DWELLING

1 Kirkland, Stephane, 'Le Grand Paris: Party 1, the launch', 25 October 2011, http://stephanekirkland.com/le-grand-paris-part-1

2 'La Ville Fertile', Cite de l'Architecture & du Patrimoinie, http://www.citechaillot.fr/fr/expositions/expositions_en_location/24047-la_ville_fertile.html

3 http://www.doorsofperception.com/infrastructure-design/life-is-a-picnic-in-the-fertile-city/

4 'Eco Boulevards', Buckminster Fuller Institute, 2010, http://challenge.bfi.org/2010Finalist_EcoBlvd

5 '55 000 hectares pour la nature', Bordeaux Métropole, 1 January 2015, http://www.lacub.fr/nature-cadre-de-vie/55-000-hectares-pour-la-nature

6 '2010–2020 La décennie bordelaise', Bordeaux Métropole, February 2012, http://www.bordeaux-metropole.fr/sites/default/files/PDF/publications/mag-Mipim-2012pdf.pdf

7 'Wild City', STEALTH.unlimited, http://www.stealth.ultd.net/stealth/01_wildcity.html

8 Neuwirth, Robert, *Stealth of Nations: The Global Rise of the Informal Economy*, New York: Anchor Books, 2012

9 Mooshammer, Helge (ed.), 'Urban Informality', Other Markets, http://www.othermarkets.org/index.php?tdid=10

10 Balmer, Kevin et al., *The Diggable City: Making urban agriculture a planning priority*, Community-Wealth, June 2005, http://www.diggablecity.org/dcp_finalreport_PSU.pdf

11 'History of NeighborSpace', City of Chicago, 11 February 2015, http://www.neighbor-space.org

12 Newton, Aaron, 'Can We Stay in the Suburbs?', The Oil Drum, 17 April 2008, http://www.theoildrum.com/node/3833

13 Holmgren, David, 'Retrofitting the Suburbs for Sustainability', *Resilience*, 5 April 2005, http://www.holmgren.com.au/DLFiles/PDFs/Holmgren-Suburbs-Retrofit-Update49.pdf

14 Green, Jared, '500 Million Reasons to Rethink the Parking Lot', Grist, 7 June 2012, http://grist.org/cities/500-million-reasons-to-rethink-the-parking-lot/

15 http://parkingday.org/resources

16 Etter, Lauren, 'Roads to Ruin: Towns rip up the pavement', *Wall Street Journal*, 17 July 2010, http://online.wsj.com/article/SB10001424052748704913304575370950363737746.html

17 Rundgren, Gunnar, 'EU Farm Land Increasingly Sealed', Garden Earth, 13 June 2011, http://gardenearth.blogspot.se/2011/06/eu-farm-land-increasingly-sealed.html

18 Bardi, Ugo, 'Getting Our Land Back', Resource Crisis, 17 April 2012, http://cassandralegacy.blogspot.fr/2012/04/getting-our-land-back.html

19 Beijer, Jorick, 'Samen de droom realiseren', Gebiedsontwikkeling, 1 October 2013, http://www.gebiedsontwikkeling.nu/artikel/6913-samen-de-droom-realiseren

20 Basu, Soma, '100 Smart Satellite Cities on Anvil', Down To Earth India, 11 July 2014, http://www.downtoearth.org.in/content/100-smart-satellite-cities-anvil

21 Warren, Rick, 'Does Your Church Really Need a Bigger Building?', Pastors, 6 February 2015, http://pastors.com/big-buildings/

22 http://crit.in/initiatives/emerging-urbanism/being-nicely-messy/

23 Daly, Herman E. and Kenneth N. Townsend, *Valuing the Earth: Economics, Ecology, Ethics*, Cambridge: MIT Press, 1993, p. 267

24 'Iowa Deconstruction & Recycling Inventory', Iowa Workforce Development, April 2011, http://www.iowaworkforce.org/greenjobs/publications/deconstruction_strategiespt3.pdf

25 'Volunteers Harvest Old Materials for Girl Scouts Eco-Camp', *Savannah Now*, 25 July 2010, http://savannahnow.com/news/2010-07-25/volunteers-harvest-old-materials-girl-scouts-eco-camp#.UEReEnjralI

26 http://www.bricksandbread.com/index/about-us-2/

27 http://atu.org.uk/

28 http://cityasbiotope.blogspot.fr

29 http://www.appearing-and-disappearing-landscapes.blogspot.com

30 http://www.mosaic-region.no

31 'U.S. Environmental Footprint', Center for Sustainable Systems, University of Michigan, October 2014, http://css.snre.umich.edu/css_doc/CSS08-08.pdf

32 'Growing Greener Cities in Africa: First status report on urban and peri-urban horticulture in Africa', Food and Agriculture Organization, 2012, http://www.fao.org/docrep/016/i3002e/i3002e.pdf

33 Grewal, Parwinder, 'Cleveland, Other Cities Could Produce Most of Their Food: Ohio State Study', Ohio Agricultural Research and Development Center, 1 September 2011, http://oardc.osu.edu/7023/Cleveland-Other-Cities-Could-Produce-Most-of-Their-Food-Ohio-State-Study.htm

34 Solomon, Debra, 'Community Pickle, a proposal', Culiblog, 30 October 2008, http://culiblog.org/2008/10/communaute-choucroute-community-picklea-proposal/

35 http://shoetowntobrewtown.com

36 http://www.wildaboutgardens.org

37 http://www.pollinatorpathway.com/about/what-is-it

38 http://www.bbc.co.uk/programmes/p00p7b36

39 Fellet, Melissae, 'Cities Can Be Carbon Sinks too, not just Sources', *New Scientist*, 13 July 2011, http://www.newscientist.com/blogs/shortsharpscience/2011/07/cities-can-be-carbon-sinks-too.html

40 Rojas-Burke, Joe, 'More Trees in a City Bring Surprising Benefit, Portland Study Finds', Oregon Live, 24 January 2011, http://www.oregonlive.com/health/index.ssf/2011/01/more_trees_in_a_city_bring_sur.html

41 http://www.treepeople.org/

42 http://www.urbanforestcoalition.com

43 http://www.fs.fed.us/ucf/news.html

44 http://www.frontlinie.nl/index.html

45 http://urbanherbology.org/tag/lynn-shore

46 http://boskoi.org/reports

47 Kinver, Mark, 'Study Shows Urbanisation's Impact on Biodiversity', BBC News & Environment, 12 February 2014, http://www.bbc.co.uk/news/science-environment-26140827

48 http://cbc.iclei.org/urbis

49 Goode, David, *Nature in Towns and Cities*, London: William Collins, 2014

50 Belt, Ken et al., 'Urban Wildlife', United States Department of Agriculture, 30 April 2014, http://www.fs.fed.us/research/wildlife-fish/themes/urban-wildlife.php

51 Xinhua, 'China's forestry developments in 2012', *China Daily*, 12 March 2013, http://www.chinadaily.com.cn/china/2013-03/12/content_16301974.htm

52 Conniff, Richard, 'Urban Nature: How to Foster Biodiversity in World's Cities', *Environment360*, 6 January 2014, http://e360.yale.edu/feature/urban_nature_how_to_foster_biodiversity_in_worlds_cities/2725/

53 Giron, Will, 'Biologists Studying NYC's Interesting Impact on Urban Wildlife Evolution', Inhabitat, 27 July 2011, http://inhabitat.com/nyc/biologists-studying-nycs-interesting-impact-on-urban-wildlife-evolution

54 Mason, Daniel, 'City of Seeds', *Lapham's Quarterly*, https://www.laphamsquarterly.org/city/city-seeds

55 http://www.urbanibalism.org/manifesto-of-urban-cannibalism

CHAPTER 5: FEEDING

1 Press Association, 'Obesity Bigger Cost for Britain than War and Terror', *The Guardian*, 20 November 2014, http://www.theguardian.com/society/2014/nov/20/obesity-bigger-cost-than-war-and-terror

2 Marvin, Simon and Will Medd, 'Fat City', *World Watch Magazine*, Vol. 18, No. 5, September/October 2005, http://www.worldwatch.org/node/583

3 Twilley, Nicola, 'Ten Landmarks of the Artificial Cryosphere', Edible Geography, 26 July 2014, http://www.ediblegeography.com/category/artificial-cryosphere/

4 http://www.monsanto.com/improvingagriculture/pages/producing-more.aspx

5 Tansey, Geoff and Tasmin Rajotte (eds), *The Future Control of Food: A Guide to International Negotiations and Rules on Intellectual Property, Biodiversity and Food Security*, Oxford: Earthscan, 2008. *See also* Martin-Prevel, Alice, 'World Bank, Listen! The "Doing Business" Approach to Agriculture Needs to End', Oakland Institute, 10 October 2014, http://www.oaklandinstitute.org/world-bank-listen

6 BSCAL, 'McKinsey Sees India As Worlds Food Factory', *Business Standard*, 24 May 1997, http://www.business-standard.com/article/specials/mckinsey-sees-india-as-worlds-food-factory-by-2005-197052401058_1.html

7 Lappe, Frances Moore et al., *World Hunger: 12 Myths*, New York: Grove Press, 1998

8 'Telling Factory Farming Fairy Tales', Grain, 7 November 2014, http://www.grain.org/article/entries/5072-telling-family-farming-fairy-tales

9 De Schutter, O., 'The Transformative Potential of the Right to Food', Report of the Special Rapporteur on the right to food to the UN Human Rights Council, A/HRC/25/57

10 Oudet, Maurice, 'La tomate: Un "produit sensible"', *Sedelan*, 7 April 2007, http://www.abcburkina.net/en/nos-dossiers/vu-au-sud-vu-du-sud/244-224-la-tomate-un-produit-sensible-qui-montre-bien-quil-ne-faut-pas-signer-trop-vite-les-ape

11 http://www.tradekey.com/products/frozen-chicken.html

12 Genowys, Ted, 'The Truth About Pork and How America Feeds Itself', Bloomberg, 5 December 2013, http://www.bloomberg.com/bw/articles/2013-12-05/food-safety-risk-as-pork-processors-face-fewer-usda-meat-inspectors

13 http://www.thefoodcommons.org

14 Grossi, Mark, 'Health Hazard: West Fresno the riskiest place to live in California', *Fresno Bee*, 16 March 2013, http://www.fresnobee.com/2013/03/16/3217239/west-fresno-the-riskiest-place.html

15 'Holy Water', PBS/POV, 13 July 2004, http://www.pbs.org/pov/thirst/special_holywater.php

16 Bedoian, Vic, 'Court Ruling a Blow to Agribusiness Mogul's Kern Water Bank', *Fresno Alliance*, 1 April 2013, http://fresnoalliance.com/wordpress/?p=9205

17 Hamblin, James, 'The Dark Side of Almond Use', *The Atlantic*, 28 August 2014, http://www.theatlantic.com/health/archive/2014/08/almonds-demon-nuts/379244/

18 Office of Communication, 'USDA Announces $78 Million Available for Local Food Enterprises', USDA, 5 September 2014, http://www.usda.gov/wps/portal/usda/usdahome?contentid=2014/05/0084.xml

19 http://www.thefoodcommons.org/project/food-commons-fresno-2/

20 Orsini, Francesco et al., *Urban Agriculture in the Developing World: A Review*, INRA and Springer-Verlag, 2013

21 Masi, Brad et al., 'Urban Agriculture in Rust Belt Cities', *Solutions*, Vol. 5, Issue 1, May 2014, pp. 44–53, http://www.thesolutionsjournal.com/node/237142

22 Viljoen, André and Katrin Bohm, *Second Nature Urban Agriculture*, London: Routledge, 2014

23 http://transitionculture.org/2012/02/20/when-the-hop-fields-come-to-town/

24 http://shoetowntobrewtown.com/about-the-organizers

25 Richard, Michael Graham, 'New Belgium Brewing Turning Wastewater to Cash', Treehugger, 26 June 2005, http://www.treehugger.com/corporate-responsibility/new-belgium-brewing-turning-wastewater-to-cash.html

26 Profita, Cassandra, 'In Chicago: Urban farm taps brewery to fuel aquaponics', OPB, 19 February 2013, http://www.opb.org/news/blog/ecotrope/chicagotheplant/

27 http://brockwell-bake.org.uk/

28 http://www.sustainweb.org/realbread/community_supported_baking

29 http://www.breadhousesnetwork.org

30 Ibid.

31 http://www.laruchequiditoui.fr

32 'No Agrobiodiversity without Peasants', Grain, March 2014, http://www.grain.org/article/entries/4911-no-agrobiodiversity-without-peasants

33 De Schutter, Olivier and Gaetan Vanloqueren, 'The New Green Revolution: How twenty-first-century science can feed the world', *Solutions*, Vol. 2, Issue 4, August 2011, pp. 33–44, http://thesolutionsjournal.org/node/971

34 Kirschenmann, Fred, 'On "Being There"', Leopold Center, Summer 2012, http://www.leopold.iastate.edu/news/leopold-letter/2012/summer/kirschenmann-being-there#sthash.fKwz07tv.dpuf

35 House of Commons Environmental Audit Committee, 'Sustainable Food', House of Commons, 30 April 2012, http://www.parliament.uk/documents/TSO-PDF/committee-reports/cmenvaud.HC879.pdf

36 'The Impacts of Sustainable Procurement', UNEP, 2012, http://www.unep.fr/scp/procurement/docsres/projectinfo/studyonimpactsofspp.pdf

37 http://www.mda.gov.br/portalmda/sites/default/files/ceazinepdf/cartilha-lt_PLANO_NACIONAL_DE_AGR-379811.pdf

38 'Farming Transitions: Pathways towards regional sustainability of agriculture in Europe', *Farm Path*, 29 March 2014, http://www.wiso.boku.ac.at/fileadmin/data/H03000/H73000/H73300/Ika/2014_FarmPath_ConceptualFramework.pdf

39 http://afsafrica.org/what-is-afsa/

40 'Mali: La Via Campesina and allies host an international agroecology forum to address food sovereignty', La Via

Campesina, 25 February 2015, http://
www.viacampesina.org/en/index.php/
main-issues-mainmenu-27/sustainable-
peasants-agriculture-mainmenu-42/1747-
mali-agroecology-is-in-our-hands-we-are-
building-it-further-together-opening-of-the-
international-agroecology-forum
41 Silici, Laura, 'Agroecology Offers FAO a
"New Window" on Agriculture', International
Institute for Environment and Development,
2 October 2014, http://www.iied.org/
agroecology-offers-fao-new-window-
agriculture
42 'Farming Transitions: Pathways towards
regional sustainability of agriculture in
Europe', *Farm Path*, 29 March 2014, http://
www.wiso.boku.ac.at/fileadmin/data/
H03000/H73000/H73300/Ika/2014_
FarmPath_ConceptualFramework.pdf
43 http://www.fordhallfarm.com
44 Vivero Pol, Jose Luis, 'Transition towards
a Food Commons regime: Re-commoning
food to crowd-feed the world', Social Science
Research Network, 13 January 2015, http://
papers.ssrn.com/sol3/papers.cfm?abstract_
id=2548928

CHAPTER 6: CLOTHING
1 http://www.garmentswithoutguilt.com
2 Fletcher, Kate, *Sustainable Fashion and
Textiles*, Oxford: Earthscan, 2008
3 Jobson, Elissa, 'Chinese firm steps up
investment in Ethiopia with "shoe city"',
The Guardian, 30 April 2013, http://www.
theguardian.com/global-development/2013/
apr/30/chinese-investment-ethiopia-shoe-
city
4 http://www.waterfootprint.
org/?page=files/home
5 'Pollution From Tanneries', YouTube, 26
February 2008, https://www.youtube.com/
watch?v=LyMY69tdd5o
6 Durai, G. and M. Rajasimman, 'Biological
Treatment of Tannery Wastewater', *Journal of
Environmental Science and Technology*, Vol. 4,
2011, pp. 1–17, http://www.scialert.net/fullte
xt/?doi=jest.2011.1.17&org=11
7 http://www.ethicalfashionforum.com/
the-issues/standards-labelling

8 Fletcher, Kate and Mathilda Tham (eds),
*Routledge Handbook of Sustainability and
Fashion*, Abingdon: Routledge, 2014
9 'Craft Of Use: The Practices of Use',
Centre for Sustainable Fashion, 11 February
2015, http://craftofuse.org/home/practices
10 http://botanicalgarden.berkeley.edu
11 https://www.cca.edu/academics/textiles/
sustainability
12 http://flaxproject.com
13 http://www.sustainablecotton.org/pages/
show/cleaner-cotton-field-program
14 http://www.fibershed.com
15 De Decker, Kris, 'Locally Farmed
Clothing: The Fibershed Project', *No Tech
Magazine*, 27 January 2012, http://www.
notechmagazine.com/2012/01/locally-
farmed-clothing-the-fibershed-project.html
16 http://www.fibershed.com/event/wool-
symposium
17 Braslow, Juliet, 'Fibershed Bringing "Farm-
Fresh" Clothing to the Region', University
of California Cooperative Extension,
Winter 2012, http://ucanr.edu/sites/
Grown_in_Marin/Grown_In_Marin_News/
Archived_Issues/Grown_in_Marin_News_
Winter_2012/Fibershed_bringing_farm-
fresh_clothing_to_the_region
18 Richards, Matt et al., 'Leather for Life',
Future Fashion White Papers, Earth Pledge
Foundation, 15 February 2015, http://www.
organicleather.com/organic_leather_white_
paper.pdf
19 Ibid.
20 http://www.leatherworkinggroup.com
21 Humphreys, Sal, 'The economies within
an online social network market: A case study
of Ravelry', ANZCA 09 annual conference,
Queensland University of Technology, 2009,
http://eprints.qut.edu.au/26455

CHAPTER 7: MOVING
1 Charette, Robert N., 'This Car Runs
on Code', IEE Spectrum, February
2009, http://www.real-programmer.
com/interesting_things/IEEE%20
SpectrumThisCarRunsOnCode.pdf
2 http://audi-urban-future-initiative.com/
initiative

3 Iseki, Hiroyuki et al., 'Thinking Outside the Bus', Access, Spring 2012, http://www.uctc.net/access/40/access40_outsidethebus.shtml

4 http://critmumbai.files.wordpress.com/2012/10/being-nicely-messy.pdf

5 http://audi-urban-future-initiative.com/blog/on-superpool

6 Varien, M., 'New Perspectives on Settlement Patterns: Sedentism and mobility in a social landscape', University of Arizona, 1997, http://ltrr.arizona.edu/content/new-perspectives-settlement-patterns-sedentism-and-mobility-social-landscape

7 Siegel, Bernard J. (ed.), *Annual Review of Anthropology*, Vol. 21, 1992, http://sspa.boisestate.edu/anthropology/files/2010/06/article14_mobility.sedentism-concepts-archaeological-meaures-and-effects.pdf

8 Whitelegg, John, 'Transport and Land Take', Eco-Logica Ltd, October 1994, http://www.eco-logica.co.uk/pdf/CPRELandTake.pdf

9 'High Speed Rail Leaves Taxpayers Saddled with Debt', *Daily Telegraph*, 6 July 2012, http://www.telegraph.co.uk/news/uknews/road-and-rail-transport/9381048/High-speed-rail-leaves-taxpayers-saddled-with-debt.html

10 Wolmar, Christian, 'What's the Point of HS2?', *London Review of Books*, 17 April 2014, http://www.lrb.co.uk/v36/n08/christian-wolmar/whats-the-point-of-hs2

11 Bradshaw, Chris, 'Valuing of Trips', Sierra Club, http://vault.sierraclub.org/sprawl/articles/trips.asp

12 Green, Jeffrey, 'Woes of Megacity Driving Signal Dawn of "Peak Car" Era', Bloomberg Business, 24 February 2014, http://www.bloomberg.com/news/articles/2014-02-24/woes-of-megacity-driving-signals-dawn-of-peak-car-era

13 http://eprints.uwe.ac.uk/23277/

14 http://changeobserver.designobserver.com/feature/a-tale-of-two-trains/21768

15 Rozycki, Christian, 'Ecology Profile of the German High-Speed Rail Passenger Transport System, ICE', *International Journal of Life Cycle Assessment*, Vol. 8, March 2003, http://link.springer.com/article/10.1007%2FBF02978431

16 Cosgrove, Christine, 'Tracking High-Speed Rail's Energy Use and Emissions', *Berkeley Transportation Letter*, Spring 2010, http://its.berkeley.edu/btl/2010/spring/HRS-life-cycle

17 Sahlins, Marshall, *Stone Age Economics*, New York: de Gruyter, 1972

18 Agger, Peder and Jesper Brandt, 'Dynamics of Small Biotopes in Danish Agricultural Landscapes', *Landscape Ecology*, Vol. 1, Issue 4, September 1988, pp. 227–40, http://www.springerlink.com/content/n2106055080v136j

19 'Appearing and Disappearing Landscapes', 11 January 2010, http://appearing-and-disappearing-landscapes.blogspot.fr/2010/01/encircling-field.html

20 http://www.biotope-city.net

21 Narayanan, Nayantara, 'Showing the Path to Other Indian Cities, Chennai Starts Pedestrianising Its Roads', Scroll, 10 November 2014, http://scroll.in/article/687775/showing-the-path-to-other-indian-cities-chennai-starts-pedestrianising-its-roads

22 Thackara, John, 'Cycle Commerce as an Ecosystem', http://www.doorsofperception.com/development-design/cycle-commerce-as-an-ecosystem

23 Thackara, John, 'Caloryville: The two-wheeled city', http://www.doorsofperception.com/notopic/caloryville-the-two-wheeled-city (this online text contains a large number of links and references that the author has not duplicated here)

24 http://www.lowtechmagazine.com/2012/09/jobs-of-the-future-cargo-cyclist.html

25 http://www.doorsofperception.com/mobility-design/cycle-commerce-the-red-blood-cells-of-a-smart-city

26 'From Income Tourism To Tourism P2P?' Revhotelution (sic), 21 December 2011, http://www.blogtrw.com/en/2011/12/from-income-tourism-to-tourism-p2p/

27 Caire, Gilles, 'Who Benefits from Holidays?', *Le Monde Diplo*, July 2012, http://

mondediplo.com/2012/07/15tourism

28 'Survey Reveals Travelers Growing Greener', TripAdvisor, 19 April 2012, http://www.tripadvisor.com/PressCenter-i5154-c1-Press_Releases.html

29 http://www.wwoofinternational.org

CHAPTER 8: CARING

I International Diabetes Federation (ed.), 'The Global Burden', *IDF Diabetes Atlas*, 4th ed., 2009, pp. 21–7

2 http://ki.se/en/about/biomedicum-laboratory-of-the-future

3 http://healthclusternet.eu

4 http://centennial.rockefellerfoundation.org/events/entry/top-trends

5 Mikkonen, Juha and Dennis Paphael, 'Social Determinants of Health: The Canadian facts', York University School of Health Policy and Management, May 2010, http://www.thecanadianfacts.org

6 Starfields, Barbara, MD, 'America's Healthcare System is the Third Leading Cause of Death', World Health Education Initiative, 26 July 2000, http://www.health-care-reform.net/causedeath.htm

7 Tainter, Joseph A., 'Complexity, Problem Solving, and Sustainable Societies', Dieoff.org, 1996, http://dieoff.org/page134.htm

8 Scannell, Jack W. et al., 'Diagnosing the Decline in Pharmaceutical R&D Efficiency', *Nature Reviews*, Vol. 11, March 2012, pp. 191–200, http://lukemuehlhauser.com/wp-content/uploads/Scannell-Diagnosing-hte-decline-in-pharmaceutical-RD-efficiency.pdf

9 Rockey, Sally and Francis Collins, 'One Nation in Support of Biomedical Research?' National Institutes of Health, 24 September 2013, http://nexus.od.nih.gov/all/2013/09/24/one-nation-in-support-of-biomedical-research/

10 Drain, Paul K. and Michael Barry, 'Fifty Years of U.S. Embargo: Cuba's health outcomes and lessons', *Science*, Vol. 328, No. 5978, 30 April 2010, pp. 572–3, http://www.sciencemag.org/content/328/5978/572

II Headey, Derek et al., 'The Other Asian Enigma: Explaining the rapid reduction of undernutrition in Bangladesh', *World Development*, Vol. 66, February 2015, pp. 749–61, http://www.sciencedirect.com/science/article/pii/S0305750X14002873

12 http://challenge-old.bfi-internal.org/application_summary/3139#

13 Sullivan, Andrew, 'How Much Do We Really Need Doctors?', The Dish, 5 June 2012, http://dish.andrewsullivan.com/2012/06/05/get-the-doctors-out-of-medicine

14 Demicco, Frederick J. and Marvin Cetron, 'Club Med', *Asia Biotech*, Vol. 10, No. 10, 2006, http://www.asiabiotech.com/publication/apbn/10/english/preserved-docs/1010/0527_0531.pdf

15 Leicester, Graham, 'We Cannot Be Healthy Alone: Food, health, and the Peckham Experiment', International Futures Forum, 27 May 2014, http://www.internationalfuturesforum.com/s/494

16 Rodriguez, Mario S. and Sheldon Cohen, 'Social Support', Carnegie Mellon University, http://www.psy.cmu.edu/~scohen/socsupchap98.pdf

17 http://www.lybba.org

18 http://curetogether.com

19 http://quantifiedself.com

20 Cooper, Charlie, 'Dementia Research: Drug firms despair of finding cure and withdraw funding after catalogue of failures', *Independent*, 17 February 2015

21 'The Importance of Social Innovation: What can older people in Wales learn from Quebec?', National Assembly for Wales, Health and Social Care Committee, 6 February 2012, http://www.senedd.assemblywales.org/documents/s500000911/Consultation%20response%20RC%2066%20-%20Wales%20Progressive%20Co-operators.html?CT=2

22 http://www.alzheimer100.co.uk

23 http://www.designcouncil.org.uk/projects/living-well-dementia-design-challenge

24 Murray, Robin, Geoff Mulgan, and Julie Caulier-Grice, 'How to Innovate: The tools for social innovation', Social Innovation Exchange, 2 February 2010, http://

socialinnovationexchange.org/sites/default/
files/event/attachments/Copy%20of%20
Generating_Social_Innovation%20v4.pdf
25 Bauwens, Michel, 'The Emergence of
Social Coops for Social Care: Italy and
beyond', P2P Foundation, 30 July 2013,
http://blog.p2pfoundation.net/the-
emergence-of-social-coops-for-social-care-
italy-and-beyond/2013/07/30
26 Conaty, Pat, 'Social Co-operatives a
Democratic Co-production Agenda for Care
Services in the UK', Co-operatives UK, 2014,
http://www.uk.coop/sites/storage/public/
downloads/social_co-operatives_report.pdf
27 Shobert, Benjamin, 'Bank on It: Is this
volunteer program the solution to China's
elder-care crisis?', Slate, 5 November 2013,
http://www.slate.com/articles/technology/
future_tense/2013/11/feng_kexiong_s_
volunteer_bank_plan_to_care_for_china_s_
elderly.html
28 Pilling, David, 'How Japan Stood Up to
Old Age', *Financial Times*, 17 January 2014,
http://www.ft.com/intl/cms/s/2/07d4c8a8-
7e45-11e3-b409-00144feabdc0.html#slide0
29 http://www.circlecentral.com
30 http://omikronproject.gr/grassroots_v1
31 http://tyze.com
32 http://my.clevelandclinic.org/childrens-
hospital/specialties-services/departments-
centers/integrative-medicine

CHAPTER 9: COMMONING

1 Hickman, Leo, 'Charity Condemns
Tourists' Use of Fresh Water in Developing
Countries', *The Guardian*, 8 July 2012,
http://www.theguardian.com/global-
development/2012/jul/08/fresh-water-
tourist-developing
2 Black, Maggie, *The No-Nonsense Guide
to International Development,* Oxford: New
Internationalist, 2007
3 Standing, Guy, *The Precariat: The New
Dangerous Class*, London: Bloomsbury, 2011
4 http://www.foreignpolicy.com/
articles/2011/10/28/black_market_global_
economy
5 Jansen, Leo, in 'A Sustainable
Development Strategy for the EU', Swedish
Society for Nature Conservation, January
2001, http://www.eeb.org/publication/2000/
Stockholm-Conf-Rep-30-1-01.pdf
6 Reijnders, Lucas, 'The Factor X Debate:
Setting targets for eco-efficiency', *Journal of
Industrial Ecology*, Vol. 2, No. 1, 1998
7 Bell, Beverly, 'Is Haïti Poor?', Other Worlds
Are Possible, 17 March 2011, http://www.
otherworldsarepossible.org/haiti-poor
8 Shanin, Teodor, 'Introduction', in A. V.
Chayanov, *The Theory of Peasant Economy*,
University of Wisconsin Press, 1986, http://
www.eng.yabloko.ru/Books/Shanin/
chayanov.html
9 'For Inspiration: Facts and figures', Global
Landscapes Forum, 2014, http://www.
landscapes.org/glf-2014/data-landscapes-
infographic-data-visualization-competition/
inspiration-facts-figures
10 Food and Agriculture Organization,
'Growing Greener Cities In Africa', 2012,
http://www.fao.org/docrep/016/i3002e/
i3002e.pdf
11 The Nubian Vault Association,
'Understand', 9 February 2015, http://www.
lavoutenubienne.org/en/nv-technique
12 Papanikolaou, George, 'Peer to Peer
Energy Production and the Social Conflicts
in the Era of "Green Development"', P2P
Foundation, 10 March 2010, http://www.
re-public.gr/en/?p=1918
13 Eurostat, 'Small and Medium-Sized
Enterprises', 20 January 2015, http://
ec.europa.eu/eurostat/statistics-explained/
index.php/Small_and_medium-sized_
enterprises
14 Steve (sic), 'Small Businesses Often
Choose Not to Grow', Small Business Lab,
30 July 2012, http://www.smallbizlabs.
com/2012/07/small-businesses-often-choose-
not-to-grow.html
15 Max-Neef, Manfred, 'Human Scale
Development', *Development Dialogue*, 1, 1989,
Dag Hammarskjold Foundation, Uppsala
16 Thiele, Leslie Paul, 'Review of Ecological
Utopias: Envisioning the Sustainable Society',
Ecology and Society, Vol. 4, No. 1, Art. 18, 4 July
2000, http://www.ecologyandsociety.org/
vol4/iss1/art18

17 Bibby, Andrew, 'Workers Occupy Plant as Spanish Co-operative Goes Under', *The Guardian*, 15 November 2013, http://www.theguardian.com/social-enterprise-network/2013/nov/15/spanish-co-op-workers-occupy-plant

18 'Open Money Manifesto', Open Money, 9 February 2015, http://www.openmoney.org/top/omanifesto.html

19 Savà, Peppe, 'An interview with Giorgio Agamben', Libcom, 10 February 2014, http://libcom.org/library/god-didnt-die-he-was-transformed-money-interview-giorgio-agamben-peppe-savà

20 Roos, Jerome, 'In Each Other We Trust: Coining alternatives to capitalism', *Roar Magazine*, 31 March 2014, http://roarmag.org/2014/03/moneylab-conference-alternative-currencies

21 Graeber, David, *Debt: The First 5,000 Years*, New York: Melville House, 2012

22 Mignolo, Walter, 'The Communal and the Decolonial', Turbulence, 9 February 2015, http://turbulence.org.uk/turbulence-5/decolonial/

23 Neuwirth, Robert, 'The Shadow Superpower', *Foreign Policy*, 28 October 2011, http://foreignpolicy.com/2011/10/28/the-shadow-superpower/

24 Mignolo, op. cit.

25 Troncoso, Stacco, 'Prehispanic 2.0 – Latin America's P2P Roots', P2P Foundation, 28 April 2014, http://blog.p2pfoundation.net/prehispanic-2-0-latin-americas-p2p-roots/2014/04/28

26 Gudynas, Eduardo, 'Buen Vivir: Today's tomorrow', *Development*, Vol. 54, No. 4, 2011, pp. 441–7, http://www.palgrave-journals.com/development/journal/v54/n4/full/dev201186a.html

27 Macmillen, Daniel, 'Latin American Progressives and Environmental Duplicity', Open Democracy, 23 October 2014, https://www.opendemocracy.net/daniel-macmillen/latin-american-progressives-and-environmental-duplicity

28 http://menemania.typepad.com/helene_finidori/2012/04/the-commons-at-the-core-of-our-next-economic-models.html

29 Bollier, David and Silke Helfrich (eds), *The Wealth of the Commons: A World Beyond Market & State*, Amherst: Levellers Press, 2014, http://wealthofthecommons.org

30 Marsh, Jason, 'Does Sharing Come Naturally to Kids?', Greater Good, 24 February 2011, http://greatergood.berkeley.edu/article/item/does_sharing_come_naturally_to_kids

31 'Scientists Confirm: File sharing is in our genes', P2P Foundation, 28 August 2008, http://www.p2p-blog.com/?itemid=833

32 Ostrom, Elinor, *Governing the Commons: The Evolution of Institutions for Collective Action*, Cambridge University Press, 1991

33 Andries, John M. and Marco A. Janssen, 'Sustaining the Commons', Arizona State University, 2013, http://sustainingthecommons.asu.edu/wp-content/uploads/2013/07/Sustaining-the-Commons-v101.pdf

34 De Angelis, Massimo, 'Reflections on Alternatives, Commons and Communities', The Commoner, Winter 2003, http://www.commoner.org.uk/deangelis06.pdf

35 Bollier and Helfrich (eds), op. cit.

36 Banks, Sophy, 'The Power of Not Doing Stuff', Transition Network, July 2003, https://www.transitionnetwork.org/blogs/rob-hopkins/2013-07/sophy-banks-power-not-doing

37 http://www.futureoffish.org

38 http://reospartners.com/team-view/63

39 Foss, Nicole, 'Finance and Food Insecurity', Automatic Earth, 3 April 2014, http://www.theautomaticearth.com/nicole-foss-finance-and-food-insecurity

40 Berry, Thomas, *The Great Work: Our Way into the Future*, New York: Broadway Books, 2000

41 Bosselman, Klaus and Prue Taylor, 'Governance for Sustainability', IUCN Environmental Policy and Law Paper No. 70, http://cmsdata.iucn.org/downloads/eplp_70_governance_for_sustainability.pdf

42 Stone, Christopher D., *Should Trees Have Standing? Law, Morality, and the Environment*, Oxford University Press, 2010

43 http://envisionspokane.org

44 Biggs, Shannon, 'Legalizing Sustainability? Santa Monica Recognizes Rights of Nature', People To People, 11 April 2013, http://www.globalexchange.org/blogs/peopletopeople/2013/04/11/legalizing-sustainability-santa-monica-recognizes-rights-of-nature

45 'Republic of Ecuador Constitution', Political Database of the Americas, 31 January 2011, http://pdba.georgetown.edu/Constitutions/Ecuador/ecuador.html

46 Universal Declaration of Rights of Mother Earth, 22 April 2010, http://therightsofnature.org/universal-declaration

47 Earth System Governance Project, http://www.earthsystemgovernance.org

48 Ito, Mumta, 'Being Nature – extending civil rights to the natural world', The Ecologist, 24 April 2014, http://www.theecologist.org/campaigning/2363662/being_nature_extending_civil_rights_to_the_natural_world.html; see also 'The Law of the Seed', http://therightsofnature.org/wp-content/uploads/The-Law-of-the-Seed.pdf

49 Roy, Arundhati, 'Confronting Empire', Ratical, 27 January 2003, http://ratical.org/ratville/CAH/AR012703.html

CHAPTER 10: KNOWING

1 Harvard School of Engineering, 'Synaptic transistor learns while it computes', 1 November 2013, http://www.seas.harvard.edu/news/2013/11/synaptic-transistor-learns-while-it-computes

2 Morton, Timothy, 'The Catastrophe Has Already Occurred', 13 July 2008, http://ecologywithoutnature.blogspot.fr/2008/07/catastrophe-has-already-occurred.html

3 http://techcrunch.com/2010/08/04/schmidt-data/

4 'IDC: Big data biz worth $16.9 billion by 2015', The Register, 3 December 2012, http://www.theregister.co.uk/2012/03/12/idc_cases_big_data/

5 Field, Anne, 'Venture Capital Flocks to the "Quantified Self"', Cisco, 3 June 2014, http://newsroom.cisco.com/feature/1425860/Venture-Capital-Flocks-to-the-Quantified-Self

6 Kaminska, Izabella, 'Welcome to the Desert of the Real', Financial Times, 9 October 2012, http://ftalphaville.ft.com/2012/10/09/1200681/welcome-to-the-desert-of-the-real-a-postmodern-economy/

7 Diamond, Jared, Collapse: How Societies Choose to Fail or Succeed, London: Penguin, 2011

8 Zerzan, John, Running on Emptiness: The Pathology of Civilization, Los Angeles: Feral House, 2002

9 Young, John et al., Coyote's Guide to Connecting with Nature, Duvall: Wilderness Awareness School, 2008

10 Haraway, Donna J., 'The Promises of Monsters', in Cultural Studies by Lawrence Grossberg, Cary Nelson, and Paula Treichler, London: Routledge, 1992, p. 328

11 Eisenstein, Charles, 'Development in the Ecological Age', Kosmos, Spring/Summer 2014, http://www.kosmosjournal.org/article/development-in-the-ecological-age

12 Berners-Lee, Mike, How Bad Are Bananas: The Carbon Footprint of Everything, Vancouver: Greystone Books, 2011

13 MacKay, David J. C., Sustainable Energy – Without the Hot Air, Cambridge: UIT, 2009

14 Heft, Harry, 'The participatory character of landscape', Open Space, http://www.openspace.eca.ac.uk/proceedings/PDF/Summary_Paper_Harry_Heft._AB_edit.W-out_trackg.pdf

15 Illich, Ivan, Deschooling Society, London: Marion Boyars, 1976

16 Sontag, Susan, Regarding the Pain of Others, London: Macmillan, 2004

17 Cunningham, Joseph, 'Is Beauty a Sense of Belonging?', Humane Pursuits, 21 November 2014, http://humanepursuits.com/beauty-belonging-desire

18 Simms, Andrew and Ruth Potts, 'The New Materialism', New Economics Foundation, 27 November 2012, https://dl.dropboxusercontent.com/u/77684614/New_Materialism_24%2011%2012.pdf

19 Morton, Timothy, The Ecological Thought, Harvard University Press, 2010

20 Bennett, Jane, Vibrant Matter, Duke University Press, 2010

21 Gratton, Peter, 'Vitalism and Life',

Philosophy in a Time of Error, 15 June 2010, http://philosophyinatimeoferror.com/2010/06/15/vitalism-and-life

22 Mensvoort, Koert van and Hendrik-Jan Grievink (eds), *Next Nature: How Art Saves the World,* Barcelona: Actar, 2014

23 Capra, Fritjof, *The Web of Life: A New Scientific Understanding of Living Systems,* New York: Anchor, 1997

24 Harding, Stephan, *Animate Earth: Science, Intuition and Gaia,* White River Junction: Chelsea Green, 2006

25 Rockwell, Teed, 'Neither Brain nor Ghost: A non-dualist alternative to the mind-brain identity theory', Cognitive Questions, 9 February 2015, http://www.cognitivequestions.org/contents.htm

26 Goffman, Ethan, 'God, Humanity, and Nature: Comparative religious views of the environment', ProQuest, December 2005, http://www.csa.com/discoveryguides/envrel/review.php

27 O'Neill, Patt, 'Glossary of Terminology of the Shamanic and Ceremonial Traditions of the Inca Medicine Lineage', Inca Glossary, 2014, http://www.incaglossary.org/v.html

28 Lumpkin, Tara M., 'Perceptual Diversity: Is Polyphasic Consciousness Necessary for Global Survival?', Bioregionalism, December 2006, http://www.bioregionalanimism.com/2006/12/is-polyphasic-consciousness-necessary.html

29 Lévi-Strauss, Claude, *The Savage Mind,* University of Chicago Press, 1966, p. 307 n. 85

30 Kropotkin, Peter, *Mutual Aid: A Factor of Evolution,* 1902; pdf at http://www.complementarycurrency.org/ccLibrary/Mutual_Aid-A_Factor_of_Evolution-Peter_Kropotkin.pdf

31 Buber, Martin, *I and Thou,* Eastford: Martino Fine Books, 2010 (1923)

32 McPherson, Guy, 'Rewilding Community Toolbox', February 2012, http://guymcpherson.com/2012/02/a-rewilding-community-toolbox

33 Morin, Edgar, *Homeland Earth: A Manifesto for the New Millennium – Advances in Systems Theory, Complexity and the Human Sciences,* New York: Hampton Press, 1999

BIBLIOGRAPHY

Bacon, Christopher M. et al., *Confronting the Coffee Crisis: Fair Trade, Sustainable Livelihood and Ecosystems in Mexico and Central America,* Cambridge, MA: MIT Press, 2008

Berger, John, *Hold Everything Dear: Dispatches on Survival and Resistance,* London: Verso, 2003

Berry, Thomas, *The Great Work: Our Way into the Future,* New York: Broadway Books, 2000

Bollier, David and Silke Helfrich (eds), *The Wealth of the Commons: A World Beyond Market & State,* Amherst: Levellers Press, 2014

Bortoft, Henri, *Taking Appearance Seriously: The Dynamic Way of Seeing,* Edinburgh: Floris Books, 2012

Buber, Martin, *I and Thou,* Eastford: Martino Fine Books, 2010 (1923)

Capra, Fritjof, *The Web of Life,* New York: Random House, 1996

Daly, Herman E. and Kenneth N. Townsend, *Valuing the Earth: Economics, Ecology, Ethics,* Cambridge, MA: MIT Press, 1993

Darwish, Leila, *Earth Repair: A Grassroots Guide to Healing Toxic and Damaged Landscapes,* Gabriola Island: New Society, 2013

Eisenstein, Charles, *Sacred Economics: Money, Gift and Society In the Age of Transition,* Berkeley: Evolver Editions, 2011

Fernández-Galiano, Luis, *Fire and Memory: On Architecture and Energy,* Cambridge, MA: MIT Press, 2000

Fleming, David, *Lean Logic: A Dictionary for the Future and How to Survive It,* 2011, http://www.leanlogic.net

Fletcher, Kate, *Sustainable Fashion and Textiles,* Oxford: Earthscan, 2008

Fletcher, Kate and Mathilda Tham (eds), *Routledge Handbook of Sustainability and Fashion,* Abingdon: Routledge, 2014

Foer, Jonathan Safran, *Eating Animals,* London: Hamish Hamilton, 2009

Gowdy, John, *Limited Wants, Unlimited Means: A Reader on Hunter-Gatherer Economics and* the Environment, Washington, DC: Island Press, 1988

Graeber, David, Debt: The First 5,000 Years, New York: Melville House, 2012

Greer, John Michael, The Long Descent: A User's Guide to the End of the Industrial Age, Gabriola Island: New Society, 2008

Harding, Stephan, Animate Earth: Science, Intuition and Gaia, White River Junction: Chelsea Green, 2006

Heinrich-Böll-Stiftung, Soil Atlas: Facts and Figures about Earth, Land and Fields, Berlin: Institute for Advanced Sustainability Studies, 2015

Hine, Dougald and Paul Kingsnorth, Dark Mountain, The Dark Mountain Project, 2010

Hopkins, Rob, The Transition Companion: Making Your Community More Resilient in Uncertain Times, London: Green Books, 2011

Illich, Ivan, Energy and Equity, London: Marion Boyars, 2000 (1974)

King, F. H., Farmers of Forty Centuries: Organic Farming in China, Korea, and Japan, Mineola: Dover Books, 2004

Klein, Naomi, This Changes Everything: Capitalism vs. the Climate, New York: Simon & Schuster, 2014

Laureano, Pietro, The Water Atlas: Traditional Knowledge to Combat Desertification, Torino: Unesco/Bollati Boringhieri editore s.r.l., 2001, https://www.box.com/s/xroo4hsq2bl190vjuu4x

Lévi-Strauss, Claude, The Savage Mind, University of Chicago Press, 1966

MacKay, David J. C., Sustainable Energy – Without the Hot Air, Cambridge, MA: UIT, 2009

Macy, Joanna, Coming Back to Life, Gabriola Island: New Society, 2014

McIntosh, Alastair, Soil and Soul: People versus Corporate Power, London: Aurum Press, 2001

Meadows, Donatela, Thinking in Systems, Oxford: Earthscan, 2009

Mignolo, Walter, The Darker Side of the Renaissance: Literacy, Territoriality, and Colonization, University of Michigan Press, 2003

Monbiot, George, Feral: Searching for Enchantment on the Frontiers of Rewilding, London: Penguin, 2013

Morin, Edgar, Homeland Earth: A Manifesto for the New Millennium, New York: Hampton Press, 1999

Morton, Timothy, The Ecological Thought, Harvard University Press, 2010

Murray, Robin, 'Co-operation in the Age of Google', Cooperatives UK, 2011

Neile, Frank, Energy: Engine of Evolution, Amsterdam: Elsevier, 2005

Neuwirth, Robert, Stealth of Nations: The Global Rise of the Informal Economy, New York: Anchor Books, 2012

Shepard, Paul, Coming Home to the Pleistocene, Washington, DC: Island Press, 2004

Silvester, Hans, Natural Fashion: Tribal Decoration from Africa, London: Thames and Hudson, 2008

Snyder, Gary, The Practice of the Wild, Berkeley: Counterpoint, 2010

Sontag, Susan, Regarding the Pain of Others, London: Macmillan, 2004

Stamets, Paul, Mycelium Running: How Mushrooms Can Help Save the World, New York: Ten Speed Press, 2005

Steele, Carolyn, Hungry City: How Food Shapes Our Lives, London: Random House, 2013

Thackara, John, In the Bubble: Designing in a Complex World, Cambridge, MA: MIT Press, 2005

Thackara, John, Wouldn't It Be Great If: The Dott 07 Manual, Design Council, 2007, http://www.doorsofperception.com/wp-content/uploads/2013/12/a-Dott07.pdf

Thayer, Robert L Jnr, LifePlace: Bioregional Thought and Practice, University of California Press, 2003

Woelfle-Erskine, Cleo et al. (eds), Dam Nation: Dispatches from the Water Underground, New York: Soft Skull Press, 2007

Zerzan, John, Against Civilization: Readings and Reflections, Los Angeles: Feral House, 2005

Further reading lists and resources are available from the Doors of Perception website, http://www.doorsofperception.com/

ACKNOWLEDGMENTS

Martin Buber taught us that all real living is meeting. Infused with that spirit, this book is the fruit of encounters and shared experiences, over the past ten years, which gave rise to the questions addressed in these pages. Some of my most memorable conversations have involved Michel Bauwens, Adélia Borges, Bryan Boyer, Jimmy Carbone, Simonetta Carbonaro, Jennifer Carlson, Orsola de Castro, Elias Cattan, Ben Cerveny, Blaine Cook, Clare Cooper, Cheryl Dahle, Kris De Decker, Paula Dib, Hugh Dubberly, Nick Durrant, Ana Dzokic, Kate Fletcher, Synnove Fredericks, Ed Gillespie, Dori Gislason, Eva Gladek, Lynda Grose, Claire Hartten, Thor Hartten, Paul Hawken, Dougald Hine, Sophia Horwitz, Jamer Hunt, Juha Huuskonen, Desiree Ibinarriaga, Alex Ioannis, Yun Im Kim, Linda Kwon, Ezio Manzini, John Manoochehri, Robin Murray, Marc Neelen, Per Olsson, Natalie Ortiz, Jogi Panghaal, Paul Polak, M. P. Ranjan, Birger Sevaldson, Aaisha Slee, Carolyn Steele, Andrew Taggart, Päivi Tahkokallio, Ben Terrett, Nicola Twilley, Stéphane Vincent, Gill Wildman, and Dilys Williams.

Many questions came into focus when I worked intensely on set-piece events such as biennials, and juries. For enabling my participation in these events, I thank David Kester, Robert O'Dowd, David Barrie, Clare Byers, Nick Devitt, Susan Hewer, Lucy Robinson, and Lauren Tan for the Designs of the Time (Dott 07) biennial in England; Elsa Francès, Nathalie Arnould, Chloé Heyraud, Marie-Haude Caraes, Gaëlle Gabillet, and Nicholas Henninger for the Design Biennial at Saint-Etienne, France; Ayush Chauhan, Babitha George, and Avinash Kumar for the Unbox Festival in Delhi; Maarten Konings, Gerbrand Bas, and Hans Robertus for the World Design Forum in Eindhoven; Kathrin diPaola, Michael Köster, and Sophie Stigliano for the Audi Urban Innovation Award; Alice Steenland, and Natalya Sverjensky, for the Axa workshop; Delight Stone, Cliff Curry, and Emiliano Gandolphi for the Curry Stone Design Prize; Joan Ubeda and Cosima Dannoritzer for the film *The Light Bulb Conspiracy*; David McConville for the Buckminster Fuller Challenge.

A good number of the people who appear in this book met first at Doors of Perception 9 in India; I thank Aditya Dev Sood and his team for their invaluable role in that memorable event. Since then our hosts and partners for Doors of Perception xskools, which succeeded the Doors conference, have been a joy and inspiration to work with, especially: Edmund Colville, who hosted the first xskool at West Lexham, England; Michael Toivio, Aija Freimane, Bo Westerlund, Martin Avila, Magnus Lindfors, Petra Lijla, Yngve Gunnarson, and Karina Vissonova for our ongoing xskools in Sweden; Elisabeth Bastian, at the Blue Mountains Cultural Centre in Katoomba; Christopher Crouch, Sally Holmes, and Brad Pettit in Fremantle; Alan Pert, Gini Lee, and Rory Hyde in Melbourne; Mansi Gupta in Kanpur; Christian Duell and Peter Hall in Brisbane; Klaus K. Loenhart and Michele Savorgnano in Venice; Donna Holford-Lovell and Dawn Campbell in Dundee; Alison Clarke and Michael Kieslinger in Vienna; Jacqueline Otten, Michael Krohn, and Karin Zindel in Zurich; Simon O'Rafferty, Frank O'Connor, and Andrew Goodman in North Wales; Terry Irwin at CMU; Andrew Polaine in Lucerne; Simona Casarotto in Treviso; Andrew Bradley, Linda Doyle, and Barry Sheahan in Dublin; Stuart Walker in Lancaster; Serge Gheldere, Joannes Vandermeulen, and Heleen van Loon in Belgium; Aldo de Jong and Abby Margolis in Barcelona; Barbara Predan in Llubljana; Nurten Meriçer, Tuna Ozkuhadar, Ayhan Ensici, Merve Titiz, and Julie Upmeyer in Istanbul; the late Dori Gislason, Soley Stefansdottir, Andri Magnasson, and Gísli Örn Bjarnhéðinsson in Iceland; Birger Sevaldson, Bjarne Ringstad, Perann Sylvia Stokke, and Jane Pernille in Norway; Peter Krogh in Denmark; Ewa

Gołebiowsk in Poland; Julius Oförsagd in Rovaniemi; Carla Mayumi, Cristina Bilsand, Camilo Belchior, and Marcelo Melo in Brazil; Oscar Salinas Flores, Jimena Acosta, Emiliano Godoy, and Desiree Ibinarriaga in Mexico; Alvin Yip in Hong Kong; Banny Banerjee at Stanford; Kati Rubinyi, Mud Baron, Rebeca Mendez, and Adam Eeuwens in Los Angeles; Edward West, Sarah Brooks, Leslie Roberts, and Kirk Bergstrom in the Bay Area; Peter Wünsch and Rachel Deller in Halifax, Nova Scotia; Ben Brangwyn, Isabel Carlisle, and Ed Mitchell in the Transition Network.

Institutional partners have been treasured oases along the way – especially the Royal College of Art, London (Paul Thompson, Jeremy Myerson, and Clare Brass); the School of Visual Arts, New York (Allan Chochinov); and the Design Museum, London, whose director Deyan Sudjic invited me to give the Puma Sustainable Design Lecture. I am grateful to the publishers of *Design Observer* – especially the late William Drenttel, and Jessica Helfand – for giving me a platform of such quality and integrity. It has been a special pleasure to work again with Thames & Hudson; Jamie Camplin commissioned my first two books an aeon ago – and his commitment to quality and clarity has been a lifelong education for this and many other writers; that standard persists in the meticulous attention given to this text by Phoebe Clapham and Jo Murray.

Among those who have been with the Doors of Perception family from its outset, way back in 1993, I thank especially Debra Solomon, who has artfully continued to connect Doors with food and design. Our software, engineering, and design maestros – Jan Jaap Spreij, Paul Jongsma, and Nique Sanders – are all masters of their craft, and treasured friends. Tony Graham and Kim Longinotto have been generous friends and hosts over many years. For keeping mind, body, and soul more or less connected, I thank my teacher Valérie Katz. My wife Kristi, and daughter Kate, have had to tolerate a preoccupied writer for far too long – but they have provided much of the motivation for getting the project done.

INDEX